Charles Darwin

CHARLES DARWIN

A Celebration of His Life and Legacy

EDITED BY
JAMES T. BRADLEY
WITH JAY LAMAR

NewSouth Books
Montgomery

NewSouth Books
105 S. Court Street
Montgomery, AL 36104

Published in the United States by NewSouth Books, a division of NewSouth, Inc.,
Montgomery, Alabama.

Library of Congress Cataloging-in-Publication Data

Charles Darwin : a celebration of his life and legacy / [edited by] James T. Bradley.

p. cm.

Includes bibliographical references and index.

ISBN-13: 978-1-58838-281-8
ISBN-10: 1-58838-281-8

1. Darwin, Charles, 1809-1882. 2. Darwin, Charles, 1809-1882. On the origin of species. 3. Darwin, Charles, 1809-1882—Influence. 4. Naturalists—England—Biography. I. Bradley, James T., Ph.D.
QH31.D2C522 2012
576.8'2092--dc23
[B]

2012007381

Design by Randall Williams

Printed in the United States of America
by Edwards Brothers Malloy

Contents

Introduction

JAY LAMAR AND JAMES T. BRADLEY

Humankind is fortunate for the life of Charles Robert Darwin (1809–82). This volume, *Darwin: A Celebration of His Life and Legacy,* commemorates the 200th anniversary of Darwin's birth and the 150th anniversary of the first edition of his most famous book, *On the Origin of Species by Means of Natural Selection, or the Preservation of Favoured Races in the Struggle for Life.* Darwin changed the world forever with his 1859 publication of that book, whose title was shortened to *On the Origin of Species* with its sixth edition in 1872. The ideas in *On the Origin of Species* reordered the biological sciences; spawned new disciplines of inquiry such as evolutionary psychology, sociobiology, and evolutionary developmental biology (evo-devo); became foundational for modern biomedical research and practice; inspired new literature and literary criticism; were grotesquely misused by 20th-century eugenicists and social Darwinists; traumatized persons with certain theological views; and continue to alter humankind's view of itself and its place in the world.

The collective contributions to the present anthology tell an interdisciplinary story of Charles Darwin the person, Darwin's work and world-altering ideas, and Darwin's legacy. This celebratory volume is the result of a happy and stimulating collaboration between the College of Liberal Arts (CLA), the College of Sciences and Mathematics (COSAM), the Caroline Marshall Draughn Center for the Arts & Humanities, and the Outreach Committee in the Department of Biological Sciences at Auburn University, a land-grant institution on the eastern plains of Alabama. During the spring of 2009, the collaborators organized and sponsored a semester-long "Darwin Celebration" for the university and the larger community. The celebration included weekly lectures on diverse aspects of Darwin's life, ideas, and legacy, and a birthday party on February 12. The lecture series included 13 speakers from four CLA departments and two COSAM departments. In addition, three world-renowned science/evolution scholars and writers visited campus to lecture and interact with undergraduate students: paleoanthropologist Richard Leakey, author and evolutionary biologist Kenneth Miller, and science journalist and author Natalie Angier. Nine speakers and four additional Auburn University scholars contributed essays for this volume. For Darwin's birthday party, Department of Biological Sciences faculty and graduate students baked and gave away more than 800 cupcakes and distributed 2,000 commemorative bookmarks to students walking between classes.

The purpose of Auburn University's Darwin Celebration was to present Darwin's ideas and their impact on diverse disciplines for general audiences in a friendly, clear, accurate, non-proselytizing way. Commemorative bookmarks with a copy of the image of Darwin and the Galapagos finches in the frieze outside the entrance to the Auburn University's Science Center Classroom were distributed to all event attendees.

We note with interest that the centennial celebration of publication of *On the Origin of Species* occurred in the same year as the famous 1959 Rede Lecture, "The Two Cultures," by British scientist and novelist C. P. Snow.[1] Snow lamented the disconnection between science and the humanities. In fact, he noted the downright hostility that often exists between the two cultures brought on by different professional languages and differing views of the human condition. Snow wrote of the optimism of scientists and contrasted it with literary artists' focus on human tragedy and loneliness. With rare exceptions, little has happened to bring Snow's two cultures amicably together during the past 50 years. Most universities still operate within a "silo system" that isolates faculty and students into geographical and intellectual spaces: natural sciences, humanities and fine arts, business, agriculture, engineering, law, and medicine. In 2009 Jerome Kagan expanded Snow's theme in his book, *The Three Cultures*, by adding the social sciences as a distinct intellectual culture.[2]

We are gratified that Snow's two cultures and Kagan's third culture are all represented by the essays gathered here under a Darwinian roof. Essay contributors are faculty members representing six academic departments at Auburn University: Biological Sciences, Foreign Languages and Literatures, Geology, History, Philosophy, and Psychology. Moreover, we are especially pleased that the volume comes from a university in the Deep South, a region known to harbor deep and strong opposition to the theory of evolution. For us, publication of *Charles Darwin: A Celebration of His Life and Legacy* feels like a springtime breeze anticipating sunshine, warmth, thaw, emergent life, and change. The editors and contributors thank Randall Williams, Suzanne La Rosa, Margaret Day, Brian Seidman, Sam Robards, Matt Johnson, Robert Carter, Lisa Harrison, Noelle Matteson, Lisa Emerson, and Jeff Benton of NewSouth Books for their hard work on this volume, including their superb editing and, most important, their willingness to make available to a wide audience these discussions of Darwin's life and work.

IN CHAPTER 1, with characteristic clarity and wit, public educator, author, and evolutionary biologist RICHARD DAWKINS leads off with an essay on the basic principles of evolution and responds to creationists' standard arguments against evolutionary theory. Dawkins's essay is a transcript of a lecture he delivered at Auburn University in 1996. While en route to Auburn, Professor Dawkins heard about the "Alabama Insert," a disclaimer of evolution by the Alabama State Board of Education pasted inside the cover of biology textbooks. Dawkins set aside his prepared lecture and extemporaneously critiqued the "Insert." Later he gave permission for a transcript of his talk to be used to further the public understanding of evolution.

CHAPTERS 2–5 DESCRIBE what Darwin did during his lifetime and give insight into what led to his theory of evolution via natural selection.

DAVID KING (Chapter 2) documents Darwin's training in geology and his little-known accomplishments as a geologist. Two great 19th-century geologists, Adam Sedgwick (1785–1873) and Charles Lyell (1797–1875), were strong and positive influences on Darwin. Cambridge's Sedgwick showed Darwin the methods of a field geologist, and his inspiring teaching countered Darwin's earlier negative experience with a poor teacher of geology at Edinburgh. Lyell authored *Principles of Geology*, of which Darwin read *Volume 1* while on the voyage of the HMS *Beagle* (1831–36). Lyell's geological uniformitarianism provided the framework Darwin needed for the vast periods of time for natural selection to produce the biological change and diversity that he observed in the fossil record and in the living world all around him. King tells how Darwin came to write four significant books on geology and then speculates about why Darwin did not pursue his love of geology after returning home from the *Beagle's* voyage.

JON ARMBRUSTER (Chapter 3) writes about the grand age of Natural History (late 1700s to about 1900), how it helped shape Darwin as a biologist, and how Darwin in turn influenced the character of the age. From this essay we learn what for some will be shocking news—that magpies were more important than the famous finches of the Galapagos Islands for Darwin's development of the concept of natural selection. Armbruster brings us up to date about the state of natural history collections worldwide and in Alabama, their value, and a recent use made of some of Darwin's specimens that the collector could never have anticipated.

GERARD ELFSTROM (Chapter 4) describes the influence that the writing of Thomas Robert Malthus (1766–1834), British demographer and political economist, had on Darwin and Darwin's contemporary, Alfred Russel Wallace (1823–1913), who independently developed a theory of evolution by natural selection. Especially interesting is how Darwin and Wallace interpreted a part of Malthus's "Principles of Population" differently in the context of the origin of human morality. Elfstrom concludes with a Malthusian analysis of current and future human growth and development of a logical and morally satisfying strategy for lowering birth rates in regions of the world least able to support greater numbers of human beings.

DEBBIE FOLKERTS (Chapter 5) tackles the controversial topic of sexual selection, competition between members of the same sex for possession of mates and/or the choosing of mates by the members of one sex. Folkerts deftly describes the controversy among biologists over whether sexual selection is distinct from natural selection or just a special case of the latter. With numerous examples, from cannibalistic spiders to colorfully gaudy bird and flower displays, Folkerts makes this little-known and poorly understood subject come alive. She convincingly argues that, as is almost always the case, Darwin's 150-year-old ideas are still on the mark.

CHAPTERS 6 AND 7 deal with two early applications of Darwin's theory of evolution by natural selection: human origins and social Darwinism. Near the end of *On the Origin of Species* Darwin cryptically remarked, "Light will be thrown on the origin of man and his history." And Darwin did just that in 1871 with publication of *The Descent of Man, and Selection in Relation to Sex.*

SHAWN JACOBSON (Chapter 6) uses the huge corpus of knowledge about human evolution and uniqueness accumulated since Darwin as a jumping-off place for informed speculation about the future evolution of *Homo sapiens*. What Darwin could not anticipate is now here: biotechnology that empowers us to shape our own evolution. Jacobsen combines his knowledge as a professional biologist and his creativity as a science fiction writer to urge reflection on how we ought to apply our newly acquired biotechnologies.

GUY BECKWITH (Chapter 7) compares the impact of Darwin's ideas on humankind to that of the Copernican Revolution. Nicolaus Copernicus (1473–1543) and Galileo Galilei (1564–1642) removed Earth and humankind from the center of the universe. Then Darwin dealt the human ego a second blow by making humankind the product of natural selection acting on chance variations, not necessarily the purposeful creation of a Creator. Beckwith details the Victorian world's strong opposition to evolution by natural selection after publication of *On the Origin of Species*. Yet Darwin's ideas not only persisted, they flourished. Why? Beckwith persuasively argues for two major factors: (1) Darwin's personal characteristics as a scientist and (2) the ease with the theory of natural selection was coopted by those with social and political agendas.

A major part of Darwin's legacy is the spawning of entirely new areas of investigation in which evolution by natural selection makes sense of otherwise relatively incomprehensible observations. Comparative and evolutionary psychology, evolutionary developmental biology (evo-devo), and origin of life studies exemplify such areas.

In the conclusion to *On the Origin of Species,* Darwin wrote: "In the distant future I see open fields for far more important researches. Psychology will be based on a new foundation, that of the necessary acquirement of each mental power and capacity by gradation." Lewis Barker (Chapter 8) and Jeffrey Katz and his co-authors (Chapter 9) chronicle Darwin's legacy in psychology.

Barker describes the pre-Darwinian philosophical roots of psychology, Darwin's forthright contributions to psychological theory, Darwin's influence on early psychologists, and the role of neo-Darwinian thinking in contemporary psychology. He argues that Darwin's *On the Origin of Species* has been more influential in psychology than the great biologist's later works that are directly related to human evolution and mental life: *The Descent of Man, and Selection in Relation to Sex* (1871) and *The Expression of Emotions in Man and Other Animals* (1872).

Comparative psychologists Katz and his co-authors acknowledge Charles Darwin as the most important figure in the creation of their discipline. Every day comparative psychologists rely on the Darwinian concept of common descent. The fact that the human brain/mind arose from ancestral nonhuman brains/minds suggests that we can gain insights about human behavior and cognitive abilities by studying the behavior and abilities of animals such as rats, pigeons, monkeys, and chimps. The question is whether the behavioral and cognitive differences between humans and modern-day nonhumans are of degree, as Darwin suggested, or of kind. Their own work with pigeons and monkeys leads Katz and his co-authors to side with Darwin.

Kenneth Halanych (Chapter 10) describes Darwin's debt to the studies of 19th century German embryologists, particularly Karl Ernst von Baer (1792–1876). He credits Von Baer's observations on animal embryos for providing Darwin with his strongest evidence for the idea of common descent. Publication of *On the Origin of Species* greatly influenced another German embryologist, Earnst Heinrich Philipp August Haeckel (1834–1919). Halanych relates how through correspondence the two scientists influenced each others' thinking about human origins and how Haeckel fanned the firestorm of controversy surrounding the theological implications of Darwin's work. The chapter concludes with the latter-20th-century revival of the importance of evolutionary theory to recent and current molecular biological studies of animal development, including emergence of the new discipline of evo-devo.

ANTHONY MOSS (Chapter 11) contributes a thorough historical account of science's attempt to answer the question of how life first emerged from inanimate matter. Beginning with the Greeks, Moss tells a captivating story about the evolution of scientists' thinking and experimentation on the origin of life question. In one of his many letters, Darwin even mused about a warm little pond on the early Earth where simple chemicals exposed to heat, light, and electricity may have begun the changes that ultimately led to life.[3] From the 1970s onward, sophisticated primitive earth "simulation" experiments and laboratory studies of molecular evolution have provided a wealth of data about life's possible origin. Darwin's principle of natural selection guides and inspires all of this work. But now, instead of a shallow, warm little pond, the evidence points to the depths of the ocean as the cradle of life. To say more would spoil this exciting, still unfolding, skillfully told tale.

FOR READERS ENJOYING chapters chronologically, GIOVANNA SUMMERFIELD's contribution on *The Adventures of Pinocchio* (Chapter 12) offers a refreshing respite from science and the history of science. Charles Darwin and Carlo Collodi (1826–90), the Florentine children's writer and creator of Pinocchio, almost certainly never met or even corresponded. Nevertheless, they had in common their free-thinking, speculative minds, perhaps nurtured by membership in the sectarian Masonry. Summerfield deftly places the struggles of the wooden puppet into the context of 19th-century

Italy's struggles for betterment as a newly united nation. Also not lost on Summerfield is the very interesting, stepwise evolution of a lifeless chunk of matter into a living, moral being.

THE LAST CHAPTER deals with society's continuing struggle with biological evolution. JAMES BRADLEY (Chapter 13) writes for would-be teachers of evolution, their students, and others with minimal knowledge about evolution but with minds willing to seek. Calling upon 35 years of experience teaching about biological evolution and origin of life studies to university students, Bradley distills the overwhelming wealth of information about the topic into what he believes all citizens should know: what evolution is, what evolution is not, and why knowing about evolution matters. He suggests using Plato's *Allegory of the Cave* as a tool for preparing students to learn about evolution.

SOME SAY THAT if Darwin had never lived, someone else would have given us the same insights about life's origins. That is surely true, for Alfred Russel Wallace, working in the same nation at the same time, came up with insights similar to Darwin's about the origin of life's diversity. But the figure we know today as the father of evolution is the iconic, white-bearded, contemplative Charles Darwin. We know about his family, which nurtured free thinking; his inspiring teachers; his voyage as a young man on HMS *Beagle*; the belated publication of his famous book; and the torment and struggles that his ideas hailed down upon him during his lifetime and bequeathed to generations following him. Darwin's story is a tale told around the world, one of scientific genius and personal perseverance. Life in the 21st century, including all of the biological sciences and their applications, is unimaginable without Darwin. From across 15 decades, Darwin invites us into his world with sheer poetry in the final sentence of *On the Origin of Species*:

> There is grandeur in this view of life, with its several powers, having been originally breathed into a few forms or into one; and that, whilst this planet has gone cycling on according to the fixed law of gravity, from

so simple a beginning endless forms most beautiful and most wonderful have been, and are being, evolved.[4] ☙

DISCUSSION GUIDES

Downloadable discussion or study guides to some of the chapters in the book, as well as links to other resources, are available at: www.newsouthbooks.com/darwin.

NOTES

1 Snow, C. 1959. The Rede Lecture. In *The Two Cultures*. Cambridge: Cambridge University Press.

2 Kagan, J. 2009. *The Three Cultures: Natural Sciences, Social Sciences, and the Humanities in the 21st Century*. New York: Cambridge University Press.

3 Darwin, F., ed. 1887. *The life and letters of Charles Darwin, including an autobiographical chapter,* vol. 3. http://darwin-online.org.uk/.

4 Darwin, C. 1859. *On the Origin of Species by Means of Natural Selection*. Mineola: Dover Publications, 2006, 307. This is an unabridged republication of the work originally published by John Murray, London, in 1859.

Charles Darwin

Amendment to the Alabama Course of Study—Science

Legislature of the State of Alabama

Text of the amendment to the Alabama Course of Study—Science, adopted by the Alabama State Board of Education in 1995, and to be pasted in all state-approved biology textbooks beginning fall, 1996:

A MESSAGE FROM THE ALABAMA STATE BOARD OF EDUCATION

This textbook discusses evolution, a controversial theory some scientists present as a scientific explanation for the origin of living things, such as plants, animals and humans.

No one was present when life first appeared on earth. Therefore, any statement about life's origins should be considered as theory, not fact.

The word "evolution" may refer to many types of change. Evolution describes changes that occur within a species. (White moths, for example, may "evolve" into gray moths.) This process is microevolution, which can be observed and described as fact. Evolution may also refer to the change of one living thing to another, such as reptiles into birds. This process, called macroevolution, has never been observed and should be considered a theory. Evolution also refers to the unproven belief that random, undirected forces produced a world of living things.

There are many unanswered questions about the origin of life which are not mentioned in your textbooks, including:

1. Why did the major groups of animals suddenly appear in the fossil record (known as the Cambrian Explosion)?

2. Why have no new major groups of living things appeared in the fossil record in a long time?

3. Why do major groups of plants and animals have no transitional forms in the fossil record?

4. How did you and all living things come to possess such a complete and complex set of "instructions" for building a living body?

5. Study hard and keep an open mind. Someday you may contribute to the theories of how living things appeared on earth.

I

The 'Alabama Insert'

A Study in Ignorance and Dishonesty

Richard Dawkins was invited in 1996 to present one of the lectures in the annual series now known as the Littleton-Franklin Lectures in Science & Humanities at Auburn University. During his visit, Professor Dawkins learned of the "Alabama Insert" and put aside his prepared text, choosing instead to deconstruct the statement by the Alabama State Board of Education. He gave permission for a transcript of his remarks and accompanying illustrations (by Lalla Ward) to be used to further evolution education in Alabama, and the lecture was published in 1997 in the Journal of the Alabama Academy of Science *68(1): 1–19. The "Alabama Insert" underwent slight revision in 2001, but because Professor Dawkins's critique of the 1995 version of the "Insert" addresses so many current misconceptions about the theory of evolution, it is republished here with permission of the journal's editor.*

Richard Dawkins

As a former prime minister of my country, Neville Chamberlain once said: "I have here a piece of paper." It says "A message from the Alabama State Board of Education." This is a flier that is designed to be—ordered to be—stuck into the front of every textbook of biology used in the public schools. What I thought I would do, with your permission, is to depart from the prepared text I brought with me. Instead I should like to go through every sentence of this document, one by one.

"This textbook discusses evolution, a controversial theory that some

> scientists present as a scientific explanation for the origin of living things such as plants, animals and humans."

This is dishonest. The use of "some scientists" suggests the existence of a substantial number of respectable scientists who do not accept evolution. In fact, the proportion of qualified scientists who do not accept evolution is tiny. A few so-called "creation scientists" are much touted as possessing PhDs, but it does not do to look too carefully where they got their PhDs from nor the subjects they got them in. They are, I think, never in relevant subjects. They are in subjects perfectly respectable in themselves, like marine engineering or chemical engineering, which have nothing to do with the matter at hand.

> "No one was present when life first appeared on Earth."

Well, that is true.

> "Therefore, any statement about life's origins should be considered as theory, not fact."

That's also true but the word *theory* is being used in a misleading way. Philosophers of science use the word *theory* for pieces of knowledge that anybody else would call fact, as well as for ideas that are little more than a hunch. It is strictly only a theory that the Earth goes around the sun. It is a theory but it's a theory supported by all the evidence. A fact is a theory that is supported by all the evidence. What this is playing upon is the ordinary language meaning of *theory* which implies something really pretty dubious or which at least will need a lot more evidence one way or another.

For example, nobody knows why the dinosaurs went extinct and there are various theories of it which are interesting and for which we hope to get evidence in the future. There's a theory that a meteorite or comet hit the Earth and indirectly caused the death of the dinosaurs. There's a theory that the dinosaurs were killed by competition from mammals. There's a theory that they were killed by viruses. There are various other theories and it is a genuinely open question which (at the time of speaking) we need more evidence to decide. That is also true of the origin of life, but it is not the case with the theory of evolution itself. Evolution is as true as the theory that the world goes around the sun.

While talking about the theories of the dinosaurs I want to make a little aside. You will sometimes see maps of the world in which the places where people speak different languages are shaded. So, you'll say, "English is spoken here," "Russian is spoken there," "French is spoken here," etc. And that's fine; that's exactly what you would expect because people speak the language of their parents.

But imagine how ridiculous it would be if you could construct a similar map for theories of, say, how the dinosaurs went extinct. Over here they all believe in the meteorite theory. Over on that continent they all believe the virus theory, down here they all believe the dinosaurs were driven extinct by the mammals. But if you think about it that's more or less exactly the situation with the world's religions.

We are all brought up with the religion of our parents, grandparents, and great-grandparents and by golly that just happens to be the one true religion. Isn't that remarkable! Creation myths themselves are numerous and varied. The creation myth that happens to be being taught to the children of Alabama is the Jewish creation myth which in turn was taken over from Babylonian creation myths and was first written down not very long ago when the Jews were in captivity. There's a tribe in West Africa that believes that the world was created from the excrement of ants. The Hindus, I am told, believe that the world was created in a cosmic butter churn. No doubt every tribe and every valley of New Guinea has its own origin myth. There is absolutely nothing special about the Jewish origin myth, which is the one we happen to have in the Christian world.

Moving on in the "Alabama Insert," as I shall call it:

> "The word 'evolution' may refer to many types of changes. Evolution describes changes that occur within a species (white moths, for example, may "evolve" into gray moths). This process is called microevolution which can be observed and described as fact. Evolution may also refer to changes of one living thing into another such as reptiles changing into birds. This process called macroevolution has never been observed and should be considered a theory."

The distinction between microevolution and macroevolution is becoming a favorite one for creationists. Actually, it's no big deal. Macroevolution is nothing more than microevolution stretched out over a much greater time span.

The moth being referred to, I presume, is the famous peppered moth, *Biston betularia*, studied in England by my late colleague Bernard Kettlewell. There is a famous story about how, in the Industrial Revolution when the trees went black from pollution, the peppered pale-colored version of this moth was eaten by birds because it was conspicuous against the black tree trunks. After the Industrial Revolution years, the black moths became by far the majority in industrial areas of England. But if you go into country areas where there is no pollution, the original peppered variety is still in a majority. I presume that's what the document is referring to.

The point about that story is that it's one of the few examples we know of genuine natural selection in action. We are not normally privileged to see natural selection in action because we don't live long enough. The Industrial Revolution, however unfortunate it may have been in other respects, did have the fortunate byproduct of changing the environment in such a way that you could study natural selection.

To study other examples of natural selection I recommend the book *The Beak of the Finch* by J. Weiner. He is describing the work of Peter and Rosemary Grant on the Galapagos finches. Those finches, perhaps more than any other animal, inspired Charles Darwin himself. What the Grants have done studying Galapagos Island finches is actually to sample populations from year to year and show that climatic changes have immediate and dramatic effects on the population ratios of various physical structures such as beak sizes.

Darwin was inspired by the example of the Galapagos finches; he was also inspired by the examples of domestication.

These are all domestic dogs [see Figure 1.1] except the top one which is a wolf. The point of it is, as observed by Darwin, how remarkable that we could go by human artificial selection from a wolf ancestor to all these breeds—a Great Dane, a bulldog, a whippet, etc. They were all produced by a process analogous to natural selection—artificial selection. Humans did the choosing, whereas in natural selection, as you know, it is nature that does the choosing. Nature selects the ones that survive and are good at reproducing, to leave their genes behind. With artificial selection, humans do the choosing of which dogs should breed and with whom they should mate.

FIGURE 1.1
The power of artificial selection to shape animals. All these domestic dogs have been bred by humans from the same wild ancestor, a wolf (top): Great Dane, English bulldog, whippet, long-haired dachshund and long-haired chihuahua.

These plants [see Figure 1.2] are all members of the same species. They are all descended quite recently from the wild cabbage *Brassica olearacea* and they are very different—cauliflower, brussels sprouts, kale, broccoli, etc. This great variety of vegetables, which look completely different, has been shaped—they have been sculpted—by the process of artificial selection from the same common ancestor.

That's an example of what can be achieved in a few centuries when the selection is powerful enough. When the selection goes on for thousands of centuries the change is going to be correspondingly greater—that's macroevolution. It's just microevolution going on for a long time.

It's difficult for the human mind to grasp how much time geology allows us, so various picturesque metaphors have been developed. The one I like is as follows: I stand with my arm outstretched and the distance from the center of my tie to my fingers represents the total time available since life began. That's about 4,000 million years. Out to about my shoulder we still get nothing but bacteria. At my elbow you might be starting to get slightly more complicated cells—eukaryotic cells—but still single cells. About mid-forearm you start getting multicellular organisms, animals you can see without a microscope. At my palm you would get the dinosaurs. Somewhere toward the end of my finger you would get the mammals. At the beginning of my nail you would get early humans. And the whole of history—all of documented written human history, all the Babylonians, Biblical history, Egyptians, the Chinese, the whole of recorded history would fall as the dust from a nail file across the tip of my furthest finger.

FIGURE 1.2

All these vegetables have been bred from the same ancestor, the wild cabbage, *Brassica olearacea*: (clockwise from top left) Brussels sprout, kohlrabi, Swedish turnip, drumhead cabbage, cauliflower and golden savoy.

This is hard for the human brain to grasp, time spans of that order. Remember that the time represented by the dust from the nail includes the time it took these cabbage varieties to evolve by artificial selection (human selection) and dogs to evolve from wolves. Just think how much change could be achieved by natural selection during the thousands of millions of years before recorded history.

To reinforce that point there was a theoretical calculation made by the great American botanical evolutionist, Ledyard Stebbins. He wanted to calculate theoretically how long it would take to evolve from a tiny mouse-sized animal (ancestor) to a descendant animal the size of an elephant. So what we are talking about is a selection pressure for increased size. Selection pressure means that in any generation slightly larger than average individuals have a slight advantage. They are slightly more likely to survive for whatever reason, slightly more likely to reproduce. Stebbins needed a number to represent that selection pressure, a way to show how strong to assume it to be. He decided to assume it (the pressure) to be so weak that you couldn't actually detect it if you were doing a field study out there trapping mice.

So Stebbins assumed his theoretical selection pressure to be so weak that it is undetectable; it vanishes in the sampling error of an ordinary research study. Nevertheless it's there. How long would it take under this small but relentless pressure for these mouse-like animals to grow and grow over the generations until they became the size of an elephant? He concluded that it would take about 20,000 generations. Well, mouse generations would be several in a year, elephant generations would take several years. Let's compromise and assume one year per generation. Even at five years per generation, that's not many years, say 100,000 years at the most. Well, 100,000 years is too short to be detected on the geological time scale for most of geologic history.

For most characteristics a selection pressure as weak as that, so weak that you couldn't even measure it, is sufficiently strong as to propel evolution so fast that it appears to be instantaneous on the geological time scale. In practice it probably isn't even as fast as that, but geological time is so vast that there is plenty of time for the evolution of all of life to have happened.

Another theoretical calculation was made by the Swedish biologist, Dan Nilsson. He took up the question which Darwin himself was interested in—the eye, the famous eye, the darling of creationist literature. Darwin himself recognized the eye as a difficult case because it is very complicated. Many people have thought, wrongly, that the eye is a difficult problem for evolutionists because—"Doesn't it have to be all there with all the bits working for the thing to work?"

No. Of course they don't all have to be there. An animal that has half

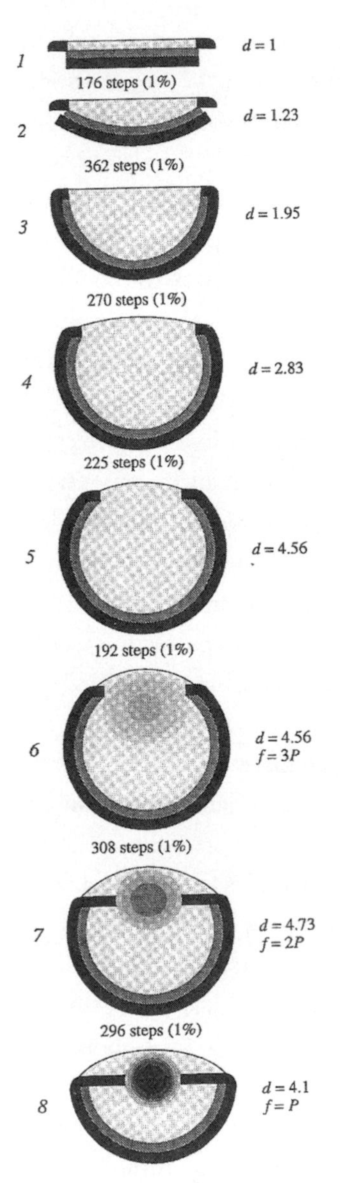

an eye can see half as well as an animal with a whole eye. An animal with a quarter eye has a quarter vision. An animal with 1⁄100 eye has 1⁄100 quality vision. It's not quite as simple as that. The point 1 am making is that you can be aided in your survival by every little tiny increment in quality of eyesight. If you have 1⁄100 quality eyesight, you can't see an image but you can see light and that might be useful. The animal might be able to tell which direction the light is coming from or which direction a shadow is coming from which could portend a predator. There are all sorts of things you could do that help you to survive if you have a small fraction of an eye, to survive better than an animal which has no eye at all. With 1⁄100 of an eye you can just about survive. With 2⁄100 of an eye you can survive a little better. There is a slow, gradual ramp of increasing probability of surviving as the eye gradually gets better.

Going back to the question of the rate at which all this happens, Nilsson did a computer modeling exercise of the evolution of the eye [see Figure 1.3]. He starts from a computer model which is not really eye shaped at all but is just a flat sheet of light sensitive cells. You've got to start somewhere.

FIGURE 1.3
Nilsson and Pelger's theoretical evolutionary series leading to a 'fish' eye. The number of steps between stages assumes, arbitrarily, that each step represents a 1 percent change in magnitude of something. See text for translation from these arbitrary units into numbers of generations of evolution.

You could start before that if you wanted to, but that's where he started. He made the computer gradually change the shapes of this model eye. The only rule was that the changes had to be small and each change had to result in an improvement in vision. The beautiful thing about the eye is that by using the actual rules of physics, the ordinary rules of optics, you can calculate how good each of the hypothetical intermediates would be at forming an image.

These intermediates all formed spontaneously in the computer as a result of gradual improvement in what the computer could measure as the optical quality of the model eye, and it goes all the way from a flat sheet of cells to a proper camera eye with a lens such as you might see in a fish. It is even better than that. The exact focusing of the lens is precisely as it should be. The details of this are written down in Nilsson's paper. By feeding in assumptions which are based upon field work in population genetics he was able to make calculations as to how long it would plausibly take under realistic conditions of natural selection. This is similar to the Stebbins calculation of how long it would take to go from the start of the series to the end.

Once again it was startlingly fast. Nilsson calculated that it would take fewer than half a million generations. The sort of small animals we are talking about, in which the eye originally evolved, would probably have had about 1 generation/year. Half a million years is a very short time on the geologic time scale.

Therefore, it's not surprising that when you look around the animal kingdom you find all the intermediates you could wish for in the evolution of the eye, in various groups of worms, etc. The eye has evolved no less than 40 times independently around the animal kingdom, and possibly as many as 60 times. So, "the" eye is really some 40 to 60 different eyes and it evolves very rapidly and exceedingly easily. There are 9 different optical principles that have been used in the design of eyes and all 9 are represented more than once in the animal kingdom.

> "Evolution also refers to the unproven belief that random, undirected forces produced a world of living things."

Where *did* this ridiculous idea come from that evolution has something to do with randomness? The theory of evolution by natural selection has a random element—mutation—but by far the most important part of the theory of evolution is non-random: natural selection. Mutation is random. Mutation is the process whereby parent genes are changed, at random. Random in the sense of not directed toward improvement. Improvement comes about through natural selection, through the survival of that minority of genes which are good at helping bodies survive and reproduce. It is the non-random natural selection we are talking about when we talk about the directing force which propels evolution in the direction of increasing complexity, increasing elegance, and increasing apparent design.

The statement that "evolution refers to the unproven belief that random undirected forces . . ." is not only unproven itself, it is stupid. No rational person could believe that random forces could produce a world of living things.

Fred Hoyle, the eminent British astronomer who is less eminent in the field of biology, has likened the theory of evolution to the following metaphor: "It's like a tornado blowing through a junkyard and having the luck to assemble a Boeing 747." His statement is a classic example of the erroneous belief that natural selection is nothing but a theory of chance. A "Boeing 747" is the end product that any theory of life must explain. The riddle for any theory to answer is, "How do you get complicated, statistically improbable apparent design?" Darwin's theory of evolution by natural selection is the only known theory that can answer this riddle. It is also supported by a great deal of evidence. With his explanation Darwin, in effect, smears out the chance or "luck" factor. There is luck in the theory, but the luck is found in small steps. Each generational step in the evolutionary process is only a little bit different from the step before. These little bits of difference are not too great to come about by chance, by mutation. However if, after the accumulation of a sufficient number of these small steps (perhaps 100), one after the other, you've got something like an eye at the end of this process, it could not have come all of a sudden by chance. Each individual step could occur by chance, but all 100 steps together could not. All 100 steps are pieced together cumulatively by natural selection.

Another metaphor along these lines is of a bank robber who went into a bank and started fiddling with the combination lock on the safe. Theoretically the thief could fiddle with the lock and have the luck to open the safe. Of course you know in practice he couldn't do that. That's why your money is safe in the bank. But just suppose that every time you twiddled that knob and got a little bit closer to the correct number, a dollar bill fell out of the safe. Then when you twiddled it another way and got a little closer still, another dollar fell out. You would very rapidly open the safe. It's like that with natural selection. Each step has a little bit of luck but when the steps are put together you end up with something that looks like a Boeing 747.

> "There are many unanswered questions about the origin of life which are not mentioned in your textbook including: why did the major groups of animals suddenly appear in the fossil record known as the "cambrian explosion."

We are very lucky to have fossils at all. After an animal dies many conditions have to be met if it is to become a fossil, and one or another of those conditions usually is not met. Personally, I would consider it an honor to be fossilized but I don't have much hope of it. If all the creatures which had ever lived had in fact been fossilized we would be wading knee deep in fossils. The world would be filled with fossils. Perhaps it is just as well that it hasn't happened that way.

Because it is particularly difficult for an animal without a hard skeleton to be fossilized, most of the fossils we find are of animals with hard skeletons—vertebrates with bones, mollusks with their shells, arthropods with their external skeleton. If the ancestors of these were all soft and then some offspring evolved a hard skeleton, the only fossilized animals would be those more recent varieties. Therefore, we expect fossils to appear suddenly in the geologic record and that's one reason groups of animals suddenly appear in the Cambrian Explosion.

There are rare instances in which the soft parts of animals are preserved as fossils. One case is the famous Burgess Shale which is one of the best beds from the Cambrian Era (between 500 million and 600 million years ago) mentioned in this quotation. What must have happened is that the ancestors of these creatures were evolving by the ordinary slow processes of evolution, but they were evolving before the Cambrian when fossilizing conditions were not very good and many of them did not have skeletons anyway. It is probably genuinely true that in the Cambrian there was a very rapid flowering of multicellular life and this may have been when a large number of the great animal phyla did evolve. If they did, their essential divergence during a period of about 10 million years is very fast. However, bearing in mind the Stebbins calculation and the Nilsson calculation, it is actually not all that fast. There is some recent evidence from molecular comparisons among modern animals which suggests that there may not have been a Cambrian Explosion at all, anyway. Modern phyla may well have their most recent common ancestors way back in the Precambrian.

As I said, we're actually lucky to have fossils at all. In any case, it is misleading to think that fossils are the most important evidence for evolution. Even if there were not a single fossil anywhere in the earth, the evidence for evolution would still be utterly overwhelming. We would be in the position of a detective who comes upon a crime after the fact. You can't see the crime being committed because it has already happened. But there is evidence lying all around. To pursue any case, most detectives and most courts of law are happy with two to three clues that point in the right direction.

Even discounting fossils, the clues that are left for us to see that prove the truth of evolution are numbered in the tens of millions. The number of clues, the sheer weight of evidence, totally and utterly, sledgehammeringly, overwhelmingly strongly supports the conclusion that evolution is true—unless you are prepared to believe the Almighty deliberately faked the evidence in order to make it look as though evolution is true. (And there are people who believe that.)

The evidence comes from comparative studies of modern animals. If you look at the millions of modern species and compare them with each other—looking at the comparative evidence of biochemistry, especially molecular evidence—you get a pattern, an exceedingly significant pattern, whereby some pairs of animals like rats and mice are very similar to each other. Other pairs of animals like rats and squirrels are a bit more different. Pairs like rats and porcupines are a bit more different still in all their characteristics. Others like rats and humans are a bit more different still, and so forth. The pattern that you see is a pattern of cousinship; that is the only way to interpret it. Some are close cousins like rats and mice; others are slightly more distant cousins (rats and porcupines) which means they have a common ancestor that lived a bit longer ago. More distinctly different cousins like rats and humans had a common ancestor who lived a bit longer ago still. Every single fact that you can find about animals is compatible with that pattern.

Similarly you can look at the geographical distribution of an animal species. Why do animals in the Galapagos Islands more closely resemble animals on neighboring islands and resemble less the animals on the mainland? It's all exactly what you would expect if evolution goes on in isolation on islands with occasional island-hopping. New foci for evolution start with migration from mainland to island and then progress from there to other islands.

If you look at the imperfections of nature you see evidence for evolution. Figure 1.4 shows animals that don't necessarily fly but are at plausible intermediate stages on the way to flight. These stages are relevant to the discussion of what's the use of half an eye or what's the use of half a wing. These animals all glide and by gliding save themselves from falling out of trees.

There are two different ways of being a flat fish. The top fish in Figure 1.5 is a skate; the bottom one is a flounder. The skate is flat the way a designer might have designed—flattened out on its belly as symmetrically as it can be. The flounder is not symmetrical because when its ancestors went flat they lay on their side, their right side. That meant that the right eye was looking down into the bottom of the sea (not good). Over many generations, natural selection favored the migration of the right eye from the underside to the top. The whole skull became distorted in an interesting way—no designer would ever have built a fish like that. The flounder has its history written all over it. Its ancestors were once free-swimming in

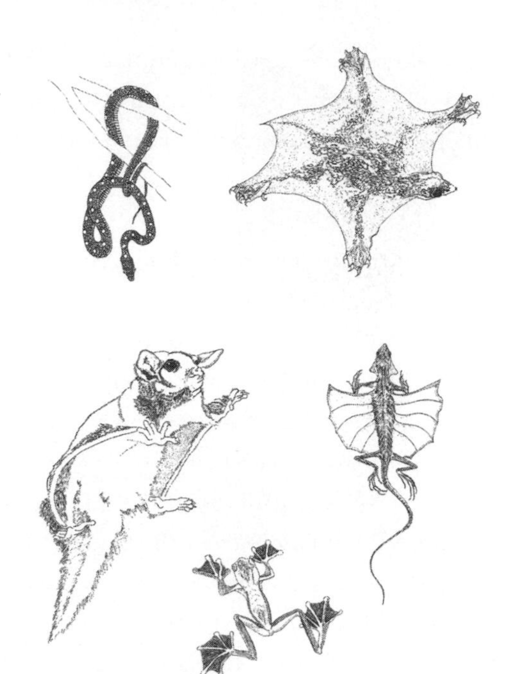

FIGURE 1.4
Vertebrates that glide down from trees but do not truly fly: (clockwise from top right) colugo, *Cynocephalus volans*; flying lizard, *Draco volans*; Wallaces flying frog, *Rhacophorus nigropalmatus*; marsupial sugar glider, *Petaurus breviceps*; and flying snake, *Chrysopelea paradisi.*

the normal way, like a trout or a salmon, and then over many generations changed into a flat fish.

> "Why have no new major groups of living things appeared in the fossil record for a long time?"

We are moving well down the list of the Alabama State Board of Education. In zoology, "major groups" would be called *phyla*—a phylum being a category such as mollusks, which includes snails and shellfish; echinoderms, which are starfish, sea urchins, and so on; chordates, which are animals with spinal cords, including ourselves; arthropods which include insects and crustaceans. The question is, "Why have no major ones appeared in a long time?"

Well, major groups don't and shouldn't, according to the Darwinian Theory, just appear. They evolve gradually. Major phyla are different from each other, though ancestrally they were like brothers. They diverged and became separate species, then separate families, then separate orders. It takes time to do that.

Think of this analogy. Suppose you have a great oak tree with huge limbs

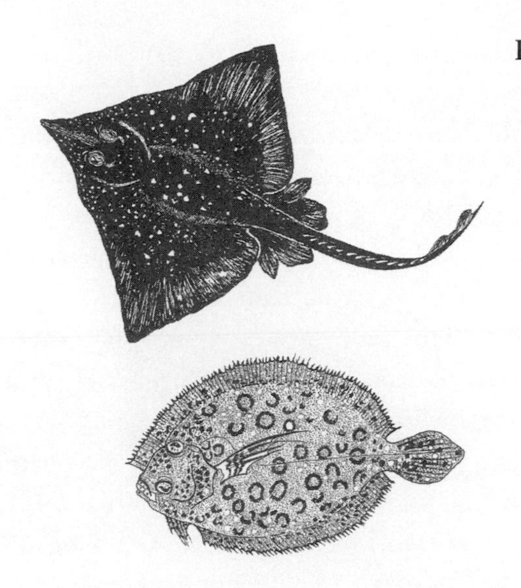

Figure 1.5
Two ways of being a flatfish: the skate, *Raja batis* (top), lies on its belly, while the flounder, *Bothus lunatus*, lies on its side.

at the base and smaller and smaller branches toward the outer layers where finally there are just lots and lots of little twigs. Obviously the little tiny twigs appeared most recently. The larger boughs appeared a long time ago and when they did appear, they were little twigs. What would you think if a gardener said, "Isn't it funny that no major boughs have appeared on this tree in recent years, only small twigs?" You'd say he is stupid.

"Why do major new groups of plants and animals have no transitional forms in the fossil record?"

It's amazing how often this is stated in the creationist literature. It's amazing because it simply isn't true. There are plenty of transitional forms. There are gaps, of course, for reasons I have stated—not all animals fossilize. But what is significant is that not a single fossil has turned up in the wrong place. Fossils are all in the right order. Creationists know that fossils all appear in the right order and it is quite an embarrassment for them. The best explanation they have come up with so far is based on Noah's flood. They say that when the great flood came the animals all rushed for the hills. The clever ones all got to the top of the hill while the stupid ones were stuck at the bottom and that's why the fossils are all neatly laid out in just the right order!

Part of the error about transitional forms may come from a misreading of a theory by my colleagues Niles Eldredge and Stephen J. Gould. Their theory is called "punctuated equilibrium." It is really about rapid gradualism or, to say it another way, gradual change that occurs rapidly separated by periods of stasis when nothing changes at all. Eldredge and Gould are rightly annoyed about the misuse of their idea by creationists, who in my terminology think punctuated equilibrium is about huge Boeing 747 type mutations. I quote Stephen Gould, "We proposed punctuated equilibrium to explain trends; it is infuriating to be quoted again and again, whether through design or stupidity I do not know, as admitting 'the fossil record includes no transition forms'. Transitional forms are generally lacking at the species level but they are abundant between larger group forms." Dr. Gould goes on, "I am both angry at and amused by the creationists and mostly I am deeply sad."

Finally, there is a semantic point about transitional forms. Zoologists, when they classify, are forced by the rules of the game to put each specimen in one species or another. In the classification business we are not allowed to say, "Well this is halfway between *Homo sapiens* and *Homo erectus*." People who dig up human fossils will always be forced to choose between one or the other. Is it *Homo erectus* or archaic *Homo sapiens*? It is forced to be one or the other. Given this definition, it is almost a legalistic point that fossils have got to be classified as one or the other. The analogy I'd offer is this. When you reach the age of majority—legal age—of 18 in Alabama you can vote. So, at the stroke of midnight on your 18th birthday you become an adult. Suppose somebody were to say, "Isn't it remarkable, there are no intermediates between children and adults?" That would be ridiculous.

"How did you and all living things come to possess such a complete and complex set of instructions for building a living body."

The set of instructions is our DNA. We got it from our parents and they got it from their parents. We can all look back through the generations, through 4,000 million years to a tiny bacterium who lived in the sea and was the ancestor of us all. We are all cousins.

We can all look back at our ancestors and claim (it's a proud claim) we are all descended from the elite. Not a single one of my ancestors died in infancy; they all reached adulthood. Not one of my ancestors failed to achieve at least one heterosexual copulation. All our ancestors were good at surviving and reproducing. We are descended from an elite.

Thousands of our ancestors' contemporaries failed. None of our ancestors did. Our DNA is DNA that has come down through thousands of millions of successful ancestors. We have inherited DNA that is pretty good at the job of surviving and, when DNA survives, it programs bodies to be good at surviving and reproducing. The world is bound to become filled with DNA that is good at surviving and reproducing. The DNA that is alive today has survived thousands of filters. Millions of generations of ancestors that survived, as a consequence of the efficient programming of their DNA, have produced an unbroken lineage. There is more to it than that. Evolution is progressive—not all the time, not uniformly—but generally it is progressive. Lineages become progressively better at what they do. Predators get better at catching prey. They have to because prey become better at getting away from predators. Just as in the human arms race there must be advances on one side to counterbalance advances on the other side.

Just a few examples of animals I would consider to be at the end of an arms race are butterflies and leaf insects (related to stick insects) that look exactly like leaves, and bugs that look like rose thorns and sit on rose stems. All of these are the result of generations of natural selection in which predators have been put off eating the ancestors of these insects. The ancestors that look most like leaves or rose thorns were the least likely to end up in predators' bellies.

The leafy sea dragon is a fish, related to sea horses. It has "fronds" that look exactly like seaweed for camouflage. This constitutes the end of an arms race in which fish that did not look like seaweed were eaten, whereas fish that did look like seaweed swam on to reproduce another day.

It's not all just survival, it's also winning mates. Birds of paradise are brightly colored because that's what females like. Genes that make pretty males are more likely to get mates and have children. This is an arms race between the salesmanship of males and the sales resistance of females.

Finally, one of the most rapid and dramatic stories of evolution—the evolution of the human brain from the brain of ape-like ancestors. The human brain constitutes the major difference between us and our close cousins, the great apes. Fossil evidence shows that our brain has blown up like a balloon during the last two or three million years as our evolution passed through the ancestral stage *Australopithecus, Homo erectus,* and finally *Homo sapiens*. No one knows why the human brain blew up in this way. I suspect again it was like some kind of arms race—some kind of positive feedback.

> "Study hard and keep an open mind. Someday you may contribute to the theories of how living things appeared on earth."

Well, at last we have found something we can agree with. This seems to me to be an admirable sentiment. I really have less trouble than some of my colleagues with so-called creation science being taught in the public schools as long as evolution is taught as well. By all means let creation science be taught in the schools. It should take all of about 10 minutes to teach it and then children can be allowed to make up their own minds in the face of evidence. For children who study hard and keep an open mind, it seems to me utterly inconceivable that they could conclude anything other than that evolution is true.

2

Darwin as a Geologist

David T. King, Jr.

Charles Robert Darwin (1809–82) is widely known as one of the greatest biological scientists of all time. What is not so commonly known is that Darwin was fascinated by the whole of the natural world and was very much interested in both the biotic and physical aspects of nature. For example, his field notes and related writings from the 1831–36 voyage of HMS *Beagle* are as much about physical features of the Earth and their origins (i.e., geology) as they are about his observations on biology.

Darwin, the son of a practicing physician, was born in Shrewsbury, Shropshire, England. He attended elementary school in Shrewsbury until 1825, when his father sent him to the University of Edinburgh to study medicine.

Edinburgh (1825–27)

At Edinburgh, Darwin's eclectic interests in the natural world began to emerge.[1] For example, he developed interests in taxidermy, marine biology, scientific collections, and note-taking about nature. He also had his first encounter with geology, but it did not go well—he took the now-infamous natural history (geology) course from Professor Robert Jameson (1774–1854).[2]

Jameson was the third Regius Professor of Natural History at Edinburgh University, a position that he held for 50 years. During his professorship he became the first strong proponent in England of the Wernerian geological system of thought, or what has been called Neptunism. Neptunism, a 19th-century philosophical approach to geology that lacked empirical data but was strong on emotion, was first developed by Abraham Gottlob Werner (1749–1817), a German mineralogy professor and eminent scholar of that time. Neptunism, which was considered to be appropriately in line with Biblical thinking of the day, held that all the Earth's rocks were formed in a single violent deluge of hot ocean water that covered the Earth and thus precipitated its mineral crust.[3]

Professor Jameson was noted for his engaging and emotional lectures, which were commonly described as spellbinding for his students. Darwin, however, was horrified by what Jameson said and his disgust with his dogmatic geology professor only grew over the academic term. When Darwin finished the course, he remarked, "I shall never take another geology class in all my life."[4]

In 1827, Darwin left Edinburgh without finishing his medical program. Darwin said that medicine was not for him. He was not fond of the sight of blood, but he was also appalled by the then-common practices of "body snatching" and murder of paupers that provided significant numbers of cadavers for the medical school at Edinburgh and elsewhere in England.[5]

CAMBRIDGE (1827–31)

Darwin's father, concerned that his son might drift into the life of an idle gentleman, arranged for him to enroll at Christ's College at Cambridge University to study for the clergy. As at Edinburgh, Darwin's eclectic interests in natural science emerged again. His interests initially focused on beetle collecting (a common pastime of that era for people interested in natural sciences), nature hikes and observations, scientific debates, and fly fishing. Later, Darwin became interested in travel for scientific investigation, mainly after learned discussions with his biological mentor at Cambridge, John Stevens Henslow (1796–1861), professor of mineralogy and botany. It was Henslow's particular method of teaching, which employed numerous field excursions and discourses in the field, that so impressed Darwin.[6]

In the summer of 1831, Darwin was introduced to Professor Adam Sedgwick (1785–1873), a clergyman and geologist who held the Woodwardian Chair at Cambridge. Sedgwick was among the most prominent geological scientists in England. His field research was helping to build the concept of geological or stratigraphic systems, which became the basis for the modern geological time scale. Sedgwick took Darwin with him to Wales during the summer of 1831 and gave him an in-depth introduction to field geology. Darwin thus saw an entirely different aspect of geology from the profoundly disappointing course that he had taken at Edinburgh.[7] In an 1831 letter to a hometown friend, Darwin said, "I am now mad about geology."[8] In another letter he stated, "I am a geologist."[9]

In 1831, Sedgwick taught Darwin the use of the field clinometer for *in situ* measurement of what Sedgwick called "dip and strike of strata," the fundamental type of measurement necessary to discern tectonic movements. Sedgwick also introduced Darwin to geologic rock classification, geologic note-taking, and geologic field interpretations.[10]

Beagle voyage (1831–36)

After finishing his clerical degree at Christ's College, Cambridge, Darwin applied for the job of naturalist aboard the HMS *Beagle*. His 1831 application was supported by Professors Henslow, Sedgwick, and others who knew of his abilities as a field naturalist.[11] On the voyage, Darwin took with him *Principles of Geology* (1830) by Charles Lyell (1797–1875), the first comprehensive textbook of geology ever written. He read this en route (1831–32) to the first stop, the Cape Verde Islands, and afterwards wrote in a letter to Henslow, "It then dawned on me that I might perhaps write a book on the geology of the various countries visited."[12]

Lyell's message to Darwin conveyed in *Principles* was this: *The Earth has great antiquity and what we see today is the result of gradual processes acting over long periods of time. Catastrophic processes, such as those held up as explanatory by the Neptunists, are not feasible to explain what we see around us today*. Darwin apparently absorbed this message deeply, and his view of the natural world was changed permanently as evidenced in his subsequent writings both geological and biological.[13]

Darwin equipped himself as a geologist for the long voyage and many potential geological discoveries. His geological toolkit included the following: geological hammer, magnifying lens, acid bottle, magnet, blowpipe, goniometer, collecting bags, specimen tags, and several sturdy field notebooks.[14]

At the *Beagle*'s first stop (January 1832), Santiago in the Cape Verde Islands, Darwin made observations over a 23-day period. His main investigations were not biological, but geological. He was intrigued by a layer of fossil shells situated 45 feet above sea level in the sea cliffs near the *Beagle*'s anchorage. Darwin noted that the shells were from ancient marine life and that their location so far above the modern sea level showed that sea level must have changed. As suggested by examples discussed in Lyell's *Principles*, Darwin interpreted this change as a gradual lowering of sea level over geological time.[15]

At the *Beagle*'s April 1832 stop at Rio de Janeiro, Brazil, Darwin trekked inland to collect biological and geological specimens, many of which were soon shipped off to Professor Henslow in England. In August 1832, the *Beagle* crew began a survey of the Patagonian coast. Darwin collected numerous Patagonian fossils, including remains of giant rodent-like animals, armadillo shells, ground sloths, many species of marine shells, and giant mammal teeth. The larger vertebrate fossils were entirely unknown to science at this time.[16] He sent these specimens to Henslow as well, along with preserved biological specimens (fish, birds, seeds, beetles, snakes, lizards, crustaceans, and plants).[17]

In 1833, after stops at Woolya Cove (Tierra del Fuego), Port Louis (Falkland Islands), and Rio Negro (Argentina), Darwin had amassed a large collection of geological specimens in addition to a trove of biological specimens. He sent these back to England, and in an accompanying July 1833 letter to Henslow he wrote, "I have endeavoured to get specimens of every variety of rock, and have written notes upon all of them." Among these specimens were samples of "quartz rock" collected on the Falklands. In his field notes, Darwin was careful to describe and sketch the rock outcrops from which he obtained specimens and where the specimens came from within the outcrops, a technique he learned from Sedgwick.[18]

In August 1833, Darwin took an overland excursion into the area of Bahia Banca, Argentina, in the company of a group of local gauchos. During this excursion, he discovered a giant vertebrate fossil, which was later determined to be an extinct ground sloth. He was fascinated by, and took extensive notes on, the simultaneous occurrence of this giant vertebrate fossil and numerous white sea shells, which reminded him of the white shell layer he had studied near Santiago. He speculated on how this fossil might have been transported from land and subsequently entombed within shelly marine sediments. Late in 1833, Darwin sent back to England another shipment of samples, which was heavily weighted toward geological and fossil specimens.[19]

During 1834–35, Darwin conducted geological and biological surveys in parts of Uruguay (Tierra del Fuego) and coastal regions of Chile, including the Chonos Archipelago.[20] According to his field notes, he was developing a particular interest in the Andes Mountains and their origins. On February 20, 1835, as Darwin was visiting Valdivia, Chile, a huge earthquake occurred. Almost every building in Valdivia was destroyed. This caused Darwin to begin thinking about tectonic motions and the origin of mountains.[21] Not long after the earthquake, Darwin was exploring the island of Quiriuina on the Chilean coast and found that the tremor had raised much of the coastline by several feet. He viewed this as direct evidence that the region of the Andes Mountains was rising gradually. To Darwin's mind, this confirmed the notions expressed in Lyell's *Principles* that small, incremental changes in sea-land level over very long time intervals could produce the great mountain ranges of the Earth. This also confirmed to Darwin the great antiquity of the Earth and the long-term nature of geological processes.[22]

Before setting sail for the Galápagos Archipelago in September 1835, Darwin organized several excursions into the Andes Mountains of Chile. He engaged in geological observations, which he recorded in his notebooks, and he collected samples. He began to make geological maps of the basic structure of the Andes, which would form the basis of his later writings and maps concerning the geology of western South America. Darwin also collected water and gas from hot springs of the Andes and sent those samples back to England (along with his rock collections) for further study.[23]

During September and October 1835, the *Beagle* crew visited several of the Galápagos islands, including Hood, James, Albemarle, Wenman, and

Culpepper (now called, respectively, Española, Santiago, Isabela, Wolf, and Darwin). Darwin is noted for his observations on the finches and tortoises of the Galápagos, but he also took extensive notes and made collections of the basaltic volcanic rocks that form the bedrock of all the Galápagos islands.[24] These samples included spectacular examples of porphyritic and vesicular basalts—basalts with large crystals (porphyroblasts) and gas bubble holes—which he collected during an extended stay on James Island.[25]

On the homeward leg of the *Beagle* voyage (1835–36), Darwin made geological and biological observations in Tahiti, New Zealand, Australia (at Sydney harbor and at Bathurst, New South Wales), Tasmania, Cocos Islands, Mauritius, Simon's Bay (near Cape Town), South Africa, St. Helena Island, and Ascencion Island. The *Beagle* arrived home in England with Darwin on board on October 2, 1836.[26]

Darwin's letters from the *Beagle*, and in particular the geological comments in those letters, were of keen interest to his friends and mentors back in England.[27] Professor Henslow was so impressed that he compiled these letters into a short book titled *Letters on Geology*, which was printed in Cambridge "for private distribution" during late 1835.[28] Darwin was unaware of this publishing effort, and when he learned of it from a letter delivered at sea, Darwin wrote to his sister: "But, as the Spaniard says '*No hay remedio*' "(there is no remedy for this).[29] The frontispiece of *Letters* contains the following statement:

> The following pages contain Extracts from Letters addressed to Professor Henslow from C. Darwin, Esq. They are printed for distribution among the Members of the Philosophical Society of Cambridge, in consequence of the interest which has been excited by some of the geological notices they contain, and which were read at a Meeting of the Society on the 16th of November 1835.[30]

POST-*BEAGLE*, GEOLOGICAL PAPERS (1836–62)

During the weeks immediately after his return, Darwin met for the first time Charles Lyell, whose book *Principles of Geology* had so profoundly affected his geological worldview. Lyell helped Darwin find appropriate naturalists with whom Darwin could leave parts of his geological collections.[31] This was an important role for Lyell because this was a time when English museums were being flooded with specimens and objects that were streaming in from colonies all around the world. One who befriended Darwin as a recipient of samples was Richard Owen (1804–92), a curator at the Hunterian Museum of the Royal College of Surgeons (later a professor and superintendent of the Natural History Department of the British Museum). Owen, who is best known for coining the term *dinosaur*, was keenly interested in Darwin's geological and fossil specimens and readily took possession of many.[32]

Not long after his return to England, Darwin was inducted as a Fellow of the Geological Society of London, at the time the only learned society for geology in the world. He was recognized for his many discoveries, observations, and collections made during the *Beagle* voyage.[33] In January 1837, Darwin gave his first post-*Beagle* paper before the Geological Society. The subject was the gradual rising of the South American continent over long spans of geological time. Darwin told the group he concluded that land masses like South America gradually rise and the nearby ocean floor subsides. Darwin also noted that these gradual changes affected habitats, and that, because of what he had observed in western South America, he felt various species would adapt to the gradual change.[34] Not long after, and simultaneously with his geological writing, Darwin began writing in his famous notebooks about his then-forming ideas on the matter of the gradual transmutation of species.[35]

In 1838, Darwin began geological investigations in England. His first and what turned out to be only set of field investigations came during an extended expedition to Glen Roy, Scotland.[36] Darwin went there to investigate the puzzling "parallel roads" phenomenon on Scottish mountain sides (Figure 2.1). The parallel roads looked to Darwin like similar features he had seen in Chile and had interpreted there as ancient shorelines.[37] Darwin's hypothesis was that the features at Glen Roy were ancient marine shorelines. Darwin laid out his evidence, but his arguments were later convincingly debunked by the famous glacial geologist Louis Agassiz (1807–73).[38] Agassiz showed how these features were instead the

FIGURE 2.1

Sketch of the "parallel roads" of Glen Roy (horizontal lines on the mountains) from Darwin's 1839 paper in the *Philosophical Transactions of the Royal Society*.[F1] Reproduced with permission from John van Wyhe, ed., *The Complete Work of Charles Darwin Online*.[F2]

result of erosion along a former glacial ice lake shoreline.[39] Later in his career, Darwin would remark that his misinterpretation at Glen Roy was an intellectual setback for him and that his being wrong bothered him through the years.[40]

By mid-1838, Darwin's health gradually worsened to the point where he was less able to do field work. His medical problems, including heart palpitations, stomach pains, nausea, headaches, and fevers, may have been caused from bites by tropical insects, but that was not known during his lifetime. Darwin suffered with these discomforts—at times quite debilitating—for the rest of his life.[41] Darwin's geological field investigations thus came to an end mainly for reasons of health. In early 1839, he married Emma Wedgewood (1808–96) and settled down to a family life in London. His continued intellectual pursuits mainly focused on the interpretation of data he had already collected.[42]

During the next few years, Darwin authored three notable geological books: *Journal of Researches into the Natural History and Geology of the Countries Visited during the Voyage of H.M.S.* Beagle;[43] *The Structure and Distribution of Coral Reefs*;[44] and *Geological Observations on South America*.[45] From *Journal of Researches*, here is a sample of Darwin's geological writing:

> I was delayed here five days, and employed myself in examining the geology of the surrounding country, which was very interesting. We here see at the bottom of the cliffs, beds containing sharks' teeth and sea-shells of extinct species, passing above into an indurated marl, and from that into the red clayey earth of the Pampas, with its calcareous concretions and the bones of terrestrial quadrupeds. This vertical section clearly tells us of a large bay of pure salt-water, gradually encroached upon, and at last converted into the bed of a muddy estuary, into which floating carcasses were swept.[46]

In *Coral Reefs* (1842), he wrote eloquently on the origin and evolution of these unique features and also included meticulously drawn maps of coral atolls of the Pacific and Indian oceans (Figure 2.2):

> A connection . . . between volcanic eruptions and contemporaneous elevations in mass has, I think, been shown to exist, in my work on coral reefs, both from the frequent presence of upraised organic remains, and from the structure of the accompanying reefs.[47]

In *South America*, Darwin included his now-famous geological cross-section of southern South America (Figure 2.3).[48] This encompassed geologic units such as "granite and andesite," "mica slate," "porphyries," "felspathic clay slate," "red sandstone," "calc(areous) slate-rock," "tuffs and ancient lavas," and "modern volcanic rocks." A contemporary published review of *South America* stated:

> It was owing to the observations of Mr. Charles Darwin, on the coast of Patagonia, that geologists were first presented with a series of

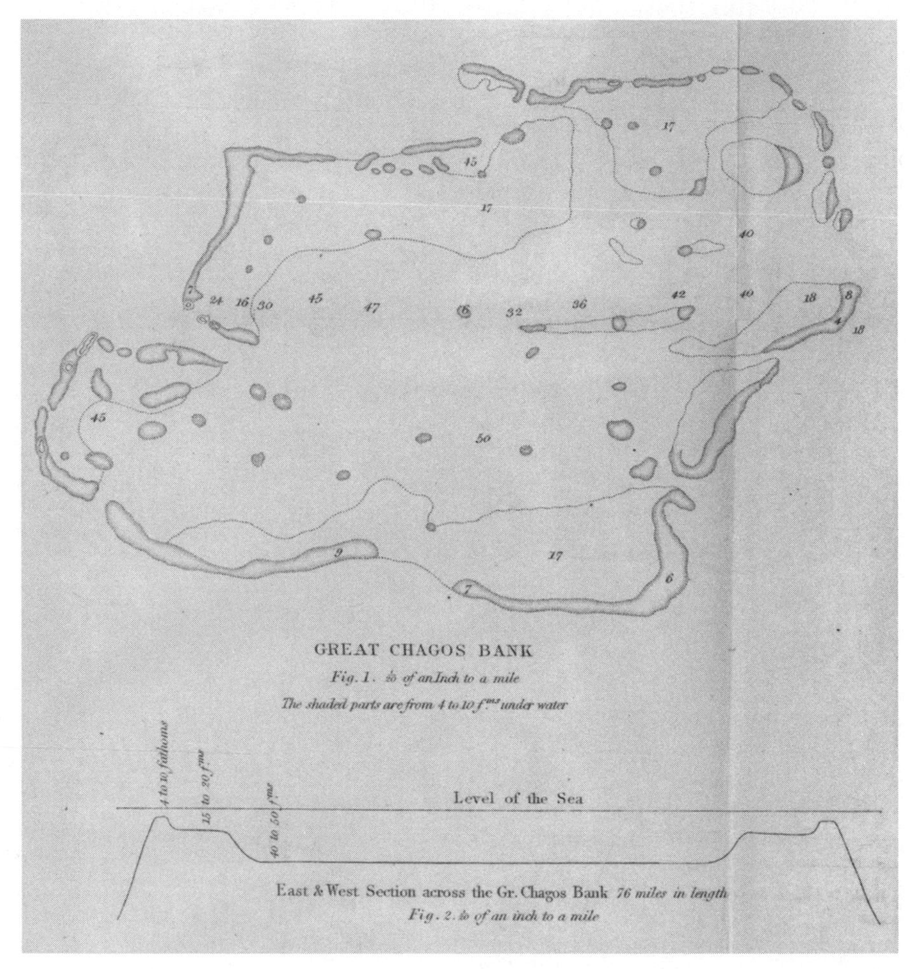

FIGURE 2.2 (above)

Example of a detailed sketch, with interpretive cross section (Great Chagos Bank atoll in the Indian Ocean [6°10'N; 72°00'E]) from Darwin's 1842 *Structure and Distribution of Coral Reefs*.[F3] Reproduced with permission from John van Wyhe, ed., *The Complete Work of Charles Darwin Online*.[F4]

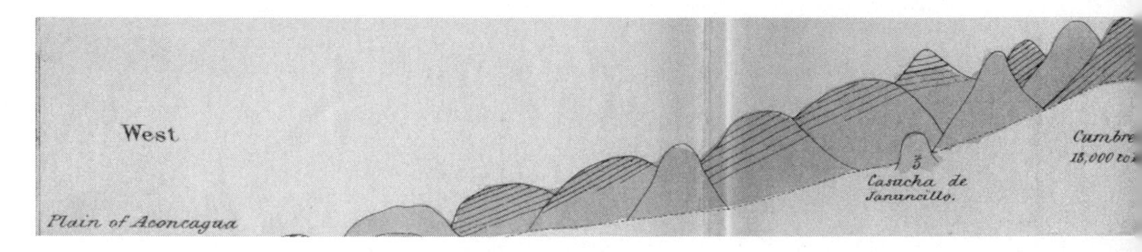

> phenomena of the gradual rising of the land, it then being in a state of repose, for a considerable period, and again rising . . . I immediately saw (how this could apply in other areas).[49]

In addition to the four notable geological works mentioned above, during the post-*Beagle* years Darwin published several other geological papers. These included seven solo-authored geological papers in the *Quarterly Journal of the Geological Society* (*QJGS*). Here is a list of those papers:[50]

1. Geological observations on the volcanic islands visited during the voyage of HMS *Beagle*, together with some brief notices on the geology of Australia and the Cape of Good Hope; being the second part of the Geology of the Voyage of the *Beagle*, under the command of Capt. Fitzroy, R.N., during the years 1832 to 1836 [*QJGS*, v. 1, p. 556–58, 1845; with a map of the Island of Ascension].
2. The structure and distribution of coral reefs; being the first part of the geology of the voyage of the *Beagle* under the command of Capt. Fitzroy, R.N., during the years 1832 to 1836 [*QJGS*, v. 1, pp. 381–89; 1845].
3. An account of the fine dust that often falls on vessels in the Atlantic Ocean [*QJGS*, v. 2, pp. 26–30; 1846].
4. On the geology of the Falkland Islands [*QJGS*, v. 2, pp. 267–74; 1846].

FIGURE 2.3 (below)
Geological cross-section from the plain of Aconcagua (Argentina) on the west (left) to Uspallata pass (Argentina) on the east—from Darwin's 1846 *Geological Observations on South America*.[F5] Reproduced with permission from John van Wyhe, ed., *The Complete Work of Charles Darwin Online*.[F6]

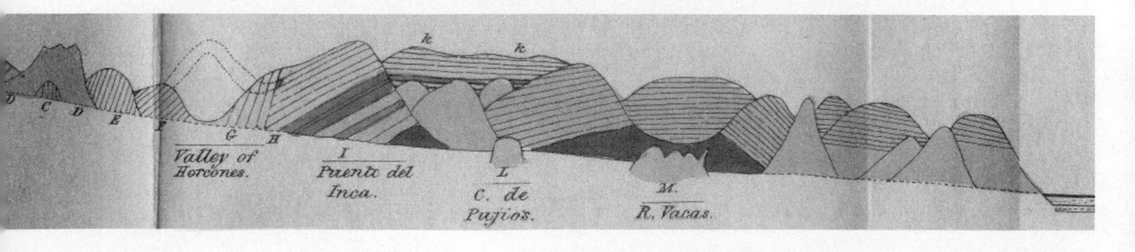

5. On the transportal of erratic boulders from a lower to a higher level [*QJGS*, v. 4, pp. 315–23; 1848].
6. On British Fossil Lepadidae [*QJGS*, v. 6, pp. 339–40; 1850].
7. On the thickness of the Pampean formation, near Buenos Ayres [*QJGS*, v. 19, pp. 68–71; 1862].

Also during this time, three Darwin-authored papers on geology were published in the *Transactions of the Geological Society of London* (*TGSL*).[51]

1. On the formation of mould [*TGSL*, ser. 2, v. 5, pt. 3, pp. 505–09; 1840].
2. On the connexion [sic] of certain volcanic phenomena in South America; and on the formation of mountain chains and volcanos [sic], as the effect of the same power by which continents are elevated [*TGSL*, ser. 2, v. 5, pt. 3, pp. 601–31; 1840].
3. On the distribution of the erratic boulders and on the contemporaneous unstratified deposits of South America [*TGSL*, ser. 2, v. 6, pt. 2, pp. 415–31; 1842].

From the time of Darwin's 1831 declaration "I am a geologist" to the end of the *Beagle* voyage in 1836, he was quite active physically and intellectually in the realm of geological thinking.[52] As noted above, he continued this geological activity to about 1838, when his ill health essentially retired him from strenuous field studies. In the publications cited above, one can see that after 1838 Darwin relied entirely on either the observations that he made during his voyage or on the Scottish field work of 1838 for his data. Darwin's 1840 paper on earthworms in soil (what he called "mould") was based on experiments in his own garden (Figure 2.4). His 1848 paper relied on observations made during the same Scottish field excursions as in his 1839 paper on the "parallel roads." Darwin's 1850 paper on Cretaceous barnacles (*Lepadidae*) appears to have been an entirely laboratory-based investigation of those fossils perhaps involving specimens given to him by others or taken from his *Beagle* collections (the paper does not say where the specimens came from). Darwin's last geological journal paper was published in 1862 (as noted above) and was based entirely upon his 1833 field work in Argentina.

One could speculate that if Darwin had not had ill health that his

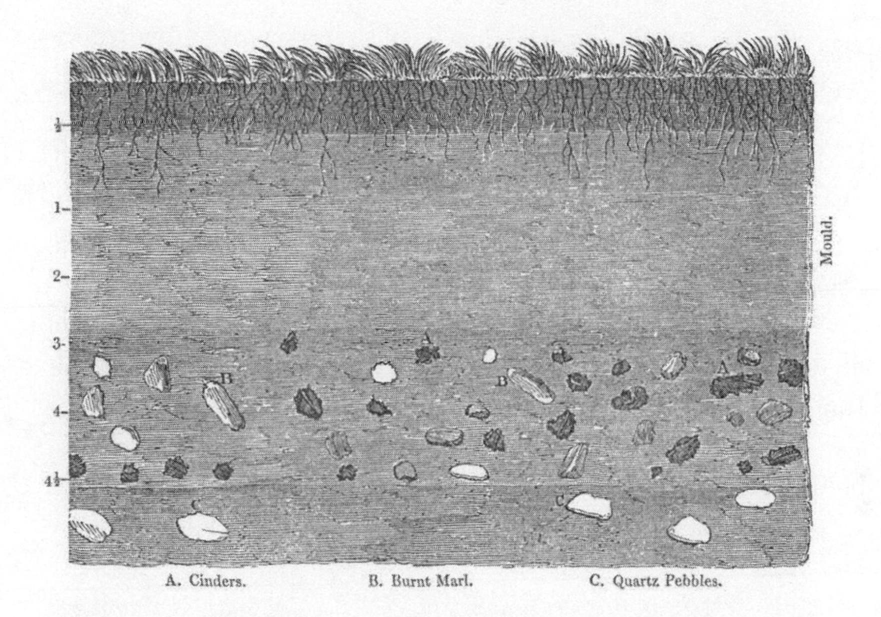

Figure 2.4
Cross-section of a soil layer where Mr. Wedgwood of Staffordshire had placed a layer of cinders, burnt marl, and quartz pebbles (listed in order of increasing density) at the surface in c. 1825, yet in 1837 Darwin observed that earthworms had churned the soil and in so doing these larger components had moved down (scale in inches is at left)—from Darwin's 1840 paper in the *Transactions of the Geological Society*. Reproduced with permission from John van Wyhe, ed., *The Complete Work of Charles Darwin Online*.[F7]

geological field work might have continued and he might have done more in the field of geology. One wonders if he might have considered returning to South America for more field studies. Whether Darwin's lingering concern about his misinterpretation of the "parallel roads" of Scotland was a factor in his turning away from geological studies in England is unclear.[53] Darwin obviously had more success in writing about South American geology. However, by the time Darwin began publishing post-1838 papers, his mind was turning increasingly toward the origin of species and other biological themes.[54]

POST-*Beagle*, BIOLOGICAL PAPERS (1843–71)

By late 1839, Darwin's first child, a son named William Erasmus Darwin, was born, and Darwin soon felt the need to move out of London to a place with more solitude and open spaces.[55] During the summer of 1842, Darwin's father purchased a home for Darwin and his family. It was called Down House, located near Kent in the southeast of England. Darwin and family moved there and lived there until his death in 1882.[56]

For Darwin, Down House was a fortress of solitude where he could do his thinking and writing as well as spend time with Emma and his family. Darwin spent progressively more time on his notebook writings about transmutation (change and the origin of species) and on other of his biological papers coming out of his *Beagle* observations. Of the books written and published during this era in Darwin's life, the following are important titles:[57]

1843—*The Zoology of the Voyage of H.M.S. Beagle Under the Command of Captain Fitzroy, R.N., during the Years 1832 to 1836* (5 volumes);

1859—*On the Origin of Species by Means of Natural Selection, or the Preservation of Favoured Races in the Struggle for Life*;

1871—*The Descent of Man, and Selection in Relation to Sex*;

1872—*The Expression of the Emotions in Man and Animals.*

During the period 1875 to 1881, he authored other biological books as well.[58]

His 1859 book, *On the Origin of Species*, is of course the most famous of all his writings. At the time of its release, a huge chorus was raised both in praise and in criticism for the book and its ideas.[59] Darwin's geological mentor, Adam Sedgwick (who was among his critics), weighed in on the book, saying in an 1859 letter to Darwin:

> If I did not think you a good tempered (and) truth loving man I should not tell you that . . . I have read your book with more pain than pleasure. Parts of it I admired greatly; parts I laughed at till my sides were almost sore; other parts I read with absolute sorrow; because I think them utterly false (and) grievously mischievous— You have deserted—after a start in that tram-road of all solid physical truth—the true method

> of induction—(and) started up a machinery as wild I think as Bishop Wilkin's locomotive that was to sail with us to the Moon. Many of your wide conclusions are based upon assumptions which can neither be proved nor disproved. Why then express them in the language & arrangements of philosophical induction?[60]

Nor did Darwin get strong support for *Origin of Species* from his other mentors. For example, Richard Owen was a strong critic and even Charles Lyell expressed concerns about some of Darwin's interpretations regarding transmutation.[61] Darwin somehow weathered the storm that this book brought upon him, his family, and supportive colleagues. From Down House, he continued to write biological tomes and papers and ponder nature during long walks in his garden until his death in April 1882.

Conclusion

Darwin as a young scientist appears to have been highly attracted to geological studies and the modes of geological thought expressed by Charles Lyell, whose influential book Darwin digested on the first leg of the *Beagle* voyage. Darwin embraced geological processes and gradualism and these concepts in turn influenced his thinking about biological change. We should recall that Darwin was inducted into the Geological Society well before he ever aligned with any other scientific group. Even though much of Darwin's work on the *Beagle* voyage was geological in nature, Darwin's geological life changed within a few years of his return to England. This may have been due in parts to the change in his health, the failure of his initial geological work in England to produce sound results, and the rise of his original fascination with the origins of the living world. For whatever reasons, Darwin's scientific career seems to be divisible into two overlapping phases, the first geological and the second biological. Most people would agree that Darwin is more famous today for his biological work and writings, but we should not forget that Darwin made impressive geological observations and interpretations, especially in South America, and once declared "I am a geologist." ☙

NOTES FOR ARTICLE

1 Herbert, S. 1986."Darwin as a Geologist." *Scientific American* 254: 94–116; 2005. *Charles Darwin, geologist*. New York: Cornell University Press.

2 Ibid.

3 Leff, D. AboutDarwin.com: Dedicated to the life and times of Charles Darwin. http://www.aboutdarwin.com/.

4 Herbert, 2005.

5 Leff; Herbert, 2005.

6 Ibid.

7 Ibid.

8 Ibid.

9 Ibid.

10 Ibid.

11 Herbert, 1986; 2005.

12 Ibid.

13 Ibid.

14 Herbert, 2005.

15 Leff; Herbert, 2005.

16 Ibid.

17 Leff.

18 Leff; Herbert, 2005.

19 Ibid.

20 Leff.

21 Ibid.

22 Leff; Herbert, 2005.

23 Ibid.

24 Herbert, 2005.

25 Leff; Herbert, 2005.

26 Ibid.

27 Ibid.

28 Darwin, C. 1835. *Letters on geology: Extracts from Letters Addressed to Professor Henslow*. Cambridge: J. S. Henslow.

29 Leff.

30 Darwin, 1835.

31 Leff.

32 Leff; Herbert, 2005.

33 Ibid.

34 Ibid.

35 Ibid.

36 Rudwick, M. 1974. "Darwin and Glen Roy: a 'Great Failure' in Scientific Meth-

od?" *Studies in History and Philosophy of Science* 5: 97–185; Leff; Herbert, 2005.

37 Leff; Herbert, 2005.

38 Darwin, C. 1839. "Observations on the Parallel Roads of Glen Roy, and of other parts of Lochaber in Scotland, with an attempt to prove that they are of marine origin." *Philosophical Transactions of the Royal Society* 129: 39–81.

39 Agassiz, L. 1967. Etudes sur les Glaciers. In *Studies on Glaciers, Preceded by the Discourse of Neuchâtel.* trans. A. Carozzi. New York: Hafner.

40 Leff; Herbert, 2005.

41 Leff.

42 Ibid.

43 Darwin, C. 1839. *Journal of Researches into the Natural History and Geology of the Countries Visited During the Voyage of H.M.S.* Beagle *Round the World.* London: John Murray.

44 Darwin, C. 1842. *The Structure and Distribution of Coral Reefs: Being the First Part of the Geology of the Voyage of the* Beagle, *Under the Command of Capt. Fitzroy, R.N. During the Years 1832 to 1836.* London: Smith Elder and Co.

45 Darwin, C. 1846. *Geological Observations on South America.* London: Smith, Elder and Co.

46 Darwin, 1839.

47 Darwin, 1842.

48 Darwin, 1846.

49 Leff.

50 van Wyhe, J. The complete works of Charles Darwin online. http://darwin-online.org.uk/.

51 Leff.

52 Herbert, 1986; 2005; Leff.

53 Rudwick, 1974.

54 Leff.

55 Ibid.

56 Ibid.

57 Ibid.

58 Ibid.

59 Ibid.

60 Ibid.

61 Ibid.

NOTES FOR FIGURES

F1 Darwin, C. 1839. "Observations on the Parallel Roads of Glen Roy, and of other parts of Lochaber in Scotland, with an attempt to prove that they are of marine origin." *Philosophical Transactions of the Royal Society* 129: 39–81.

F2 van Wyhe, J. The complete works of Charles Darwin online. http://darwin-online.org.uk/.

F3 Darwin, C. 1842. *The Structure and Distribution of Coral Reefs: Being the First Part of the Geology of the Voyage of the* Beagle, *Under the Command of Capt. Fitzroy, R.N. During the Years 1832 to 1836.* London: Smith Elder and Co.

F4 Van Wyhe.

F5 Darwin, C. 1846. *Geological Observations on South America.* London: Smith, Elder and Co.

F6 Van Wyhe.

F7 Ibid.

3

Darwin and Collections

JONATHAN ARMBRUSTER

Charles Darwin was a collector. From the time of his childhood to his old age, Darwin made collections of biological and geological materials. Even his pet pigeons found their way into museums as stuffed specimens or skeletons. This was not unusual as the natural sciences at the time were largely collection-based. The late 1700s through around 1900 saw the formation of the world's greatest natural history museums as scientists searched the world for specimens. It was deemed so important a task that even the British Navy was enlisted, with a ship's surgeon often serving as the ship's naturalist, and even some of the captains, like Captain Robert M. Fitzroy, also collecting specimens from their journeys.[1]

Darwin helped shape this grand age of natural history. Natural history had been around as a science since the Greeks, and Aristotle is often regarded as its founder, but natural history certainly goes further back than that. We are products of our environment, and have been learning about it for our entire existence. We give names to those things that are important to us, and we learn how they live so that we might use them as food or other products.

Despite its importance to the Greeks, natural history wilted in Europe in the Middle Ages. However, it flourished in the Arabic world. Even natural selection may have been presaged by the Arabic author Al-Jahiz (Amr ibn Bahr al-Kinani al-Fuqaimi al-Basri al-Jahiz, 781–868),[2] although this may be overstated.[3] Al-Jahiz in his *Book of Animals* stated: "Animals engage in a struggle for existence; for resources, to avoid being eaten and to breed. Environmental factors influence organisms to develop new characteristics to ensure survival, thus transforming into new species. Animals that survive to breed can pass on their successful characteristics to offspring."[4]

In Europe, natural history mainly muddled along through the Middle Ages following largely from Aristotle, but the explosion of new species being discovered with the advent of transoceanic voyages demanded that rules of classification be developed. The Swedish botanist Carl Linne (Carolus Linneaus, 1707–78) gave us the Linnean Classification scheme starting with *Systema Naturae*,[5] and, most importantly, he gave us the binomial system in which species are recognized as a combination of a genus and a specific epithet, and this ushered in the modern taxonomic era into which Darwin was born.

PEDIGREE

Charles Darwin was the son of Robert Darwin (1766–1848) and grandson of Erasmus Darwin (1731–1802). Erasmus was a physician, poet, and natural philosopher, among many other attributes. In the popular *The Loves of Plants*, he explains the Linnean Classification scheme, and in *Zoönomia*, he presaged the views of Lamarck that animals could evolve by selective use and disuse of characters and that these traits could be passed on to their offspring.[6]

Robert Waring Darwin also became a physician, but he was not a naturalist. In many ways he was the exact opposite of Charles. He was a successful student, doctor,[7] and financier.[8] He was also a very large man who stopped weighing himself when he was 24 stone (about 336 pounds), whereas Charles was very athletic. Both Robert and Erasmus were among the liberal elite and known for their support of the abolition of slavery. Robert was a freethinker[9] who believed that science and reason—not the supernatural—should govern life. Clearly, Robert's belief in free thought affected Charles's later use of science and reason to explain how life evolved.[10]

Charles's mother, Susannah Wedgwood, was a Unitarian, and Darwin's first school was a day school run by the Unitarians,[11] who are also known for non-dogmatic thinking. This pedigree of freethinkers, liberals, and scientists likely allowed Darwin to help reform a largely observational natural history into a true science.

SCHOOLING

Charles was sent in 1825, at the age of 16, to the medical school at the University of Edinburgh. His father wished him to attend that school because he had and because it had a radical faculty in a relatively liberal city— in other words, people more like him. Charles hated classes, even zoology, but he and his brother (Erasmus) checked out more books from the library than all other students combined, so he wasn't wasting his time. He hated the two surgeries he witnessed, one done on a child without anesthesia. He could not stand the sight of blood and had ethical problems with how human cadavers came to be dissected at the school. Clearly he was not meant to be a doctor.

In that first year in school, he learned some taxidermy, which would be essential for his future work. On January 29, 1826, he wrote to his sister about his teacher, John Edmonstone, a freed slave from British Guiana (now Guyana):[12]

> I am going to learn to stuff birds, from a blackamoor I believe an old servant of Dr. Duncan: it has the recommendation of cheapness, if it has nothing else, as he only charges one guinea, for an hour every day for two months.

Edmonstone had learned taxidermy from the famed taxidermist Charles Waterton, and he taught Charles not only how to stuff birds but also about life in the jungle and about life as a slave. These discussions made Charles long for trips into the jungle.[13] Even more important, his discussions with Edmonstone, combined with the abolitionist beliefs passed down through his family, likely were responsible for Charles's early thoughts on the common descent of humans. In other words, learning some of the basic techniques of taxonomy and the building of collections were instrumental in the formation of one of Darwin's great ideas, that of the common descent of all species in general and our own in particular.

In Edinburgh, Darwin also learned a lot from Robert Grant, an early evolutionist and a fan of Charles's grandfather. Charles became one of Grant's favorite students, in part because of his ancestry. Grant taught

Charles about marine invertebrates, which would become a lifelong fascination. Grant also taught Charles about the concept of homology.[14] Homology can be defined today as the shared evolutionary origin of a structure. In evolutionary biology, homology refers to characteristics that are similar because of shared ancestry, though the characteristic might not be used in the same way. For example the bones in the arm of a human and that of a bird are homologous, despite the fact that birds and humans use those bones quite differently. However, the presence of homologous structures can indicate a pattern of common descent.

After two years, it was clear that Darwin was failing as a student. He didn't care for his lectures and was only interested in natural history. His father famously scolded him by saying, "You care for nothing but shooting, dogs and rat catching, and you will be a disgrace to yourself and all your family."[15] Even today, this is often the belief of parents whose children wish to go into the natural sciences, and many of us can relate similar if not quite as extreme stories about our own parents.

Robert took Charles out of the University of Edinburgh and enrolled him at Christ's College, Cambridge. Now Charles was to become a country parson after completing his degree in divinity. That clearly never happened, but Darwin was a relatively good student at Cambridge, though he was still bored by his classes—he frequently did not attend them—and he still cared more for natural history.

At Cambridge, Charles met the botanist John S. Henslow. He and Henslow became so inseparable that people came to refer to Charles as "The Man Who Walks with Henslow."[16] Henslow was unique among botanists at the time in that he arranged his herbarium by collation.[17] This means that he placed several specimens on one herbarium sheet to show the range of variation of a species. This was a unique way of looking at things because variation was often not referred to when a species was defined solely by the type specimen (the typological species concept). Thus, Henslow introduced Charles to variation. Charles also contributed to the Henslow Herbarium, and his earliest known herbarium specimen (1831) is that of the sea stock (*Matthiola sinuata*; Figure 3.1).[18]

While at Cambridge, Charles also renewed his interest in collecting

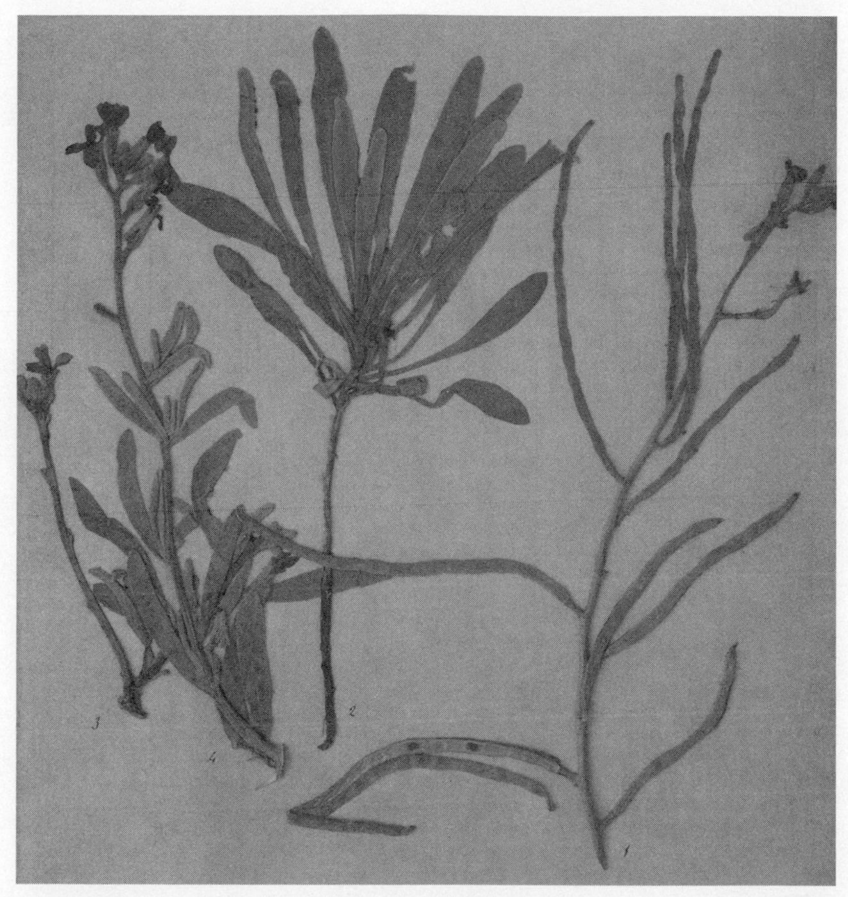

Figure 3.1
Collated herbarium sheet of three specimens of Sea Stock, *Matthiola sinuate*, collected by Charles Darwin. From Kohn, et al. (2005).

beetles. His prowess and interest in collecting was well-known by his peers. He got into a heated rivalry with fellow student Charles Babbington, who was often referred to by the moniker "Beetles." Another student, Albert Way, drew a cartoon of Charles hunting for beetles while riding a very large one. Of this period, Darwin writes about one particular episode that tells of the joys, excitement, and frustration of a collector.

> One day, on tearing off some old bark, I saw two rare beetles, and

> seized one in each hand; then I saw a third and new kind, which I could not bear to lose, so that I popped the one which I held in my right hand into my mouth. Alas! It ejected some intensely acrid fluid, which burnt my tongue so that I was forced to spit the beetle out, which was lost, as was the third one.[19]

Charles graduated from Christ's College in 1831, 10th of 171 students. He still did not know what to do with his life, but he was encouraged by Henslow to take a trip to North Wales with the geologist Adam Sedgwick. It was on this trip that he learned most of his skills in geology. It was also during this trip that Captain Robert Fitzroy once again came looking to Henslow for a traveling companion on what was supposed to be a two-year voyage on HMS *Beagle*. Henslow had a young family and had previously suggested his brother-in-law, Leonard Jenyns, as a voyager. Jenyns backed out, and this time Henslow enthusiastically suggested Charles Darwin.[20]

The position was unpaid. Contrary to common lore, Darwin was not the ship's naturalist, which was a title held by the ship's surgeon. Instead, his position was ship's gentleman, and he was basically hired on to be Captain Fitzroy's friend. He took the position because he could do natural history.[21] Throughout what would become a five-year voyage, Charles would send specimens back to Henslow and Henslow would forward them to experts for examination.

The primary mission of the HMS *Beagle* was to map southern South America, but the voyage would include many stops along the way, each providing a different habitat for Charles to explore. It was South America where the *Beagle* spent most of its time, and it was South America that really began to reveal the story of evolution to Charles.

His first stop in South America was Bahia, Brazil, and the majesty of the neotropical rainforest was immediately apparent to him.

> Delight itself, however, is a weak term to express the feelings of a naturalist who for the first time has wandered by himself in a Brazilian forest. The elegance of the grasses, the novelty of the parasitical plants, the beauty of the flowers, the glossy green of the foliage, but above all

the general luxuriance of the vegetation, filled me with admiration.[22]

His first real taxonomic work was in the Atlantic Coastal rainforest of Rio de Janeiro, where he spent three months. Among his discoveries were terrestrial flatworms, many of which he would eventually describe. Darwin speaks of the rain forest with the reverence of a natural historian when he comes to the realization that almost all was new, and all was of great interest. To a person with broad interests, the rain forest presents a unique challenge of what to collect.

> In England any person fond of natural history enjoys in his walks a great advantage, by always having something to attract his attention; but in these fertile climates, teeming with life, the attractions are so numerous, that he is scarcely able to walk at all.[23]

Most of the time that Charles spent in South America was based out of Uruguay. He spent two years exploring Uruguay and Argentina, collecting specimens, and observing the life history of plants and animals. His collections from Patagonia famously included some very important fossils (particularly mammals), such as the first fossil of the South American ungulate *Toxodon platensis*.[24]

While all of his collections were certainly important in the formation of Darwin's ideas, none are more synonymous with Charles Darwin than his collections from the Galapagos Islands, Ecuador. The *Beagle* arrived on Chatham Island (now Isla San Cristóbal), on September 17, 1835. Apparently the first specimen he procured was a specimen of the jewel moray. Although the original specimen is lost, the species would be described from his other specimens as *Muraena lentiginosa.*[25] He also observed the iguanas of the Galapagos and wrote of the terrestrial one in not so glowing, but incredibly accurate, prose.

> . . . ugly animals, of a yellowish orange beneath, and of a brownish-red colour above: from their low facial angle they have a singularly stupid appearance.[26]

Although Darwin provided a nice picture of the plants and animals of the Galapagos, it was the finches of the islands that came to bear his name. The importance of these finches to Darwin's thinking was probably pretty limited. He made a couple of crucial mistakes that were later pointed out to him by the ornithologist John Gould. First, he thought that the species were from different groups and not descended from a common ancestor. Second, he committed one of the most grievous of errors for a collector: he didn't bother to label the island of origin for many of the specimens.[27] Much of what Gould was able to determine of the so-called Darwin's finches was based on specimens that were collected by Robert Fitzroy, or on localities divined by Darwin trying to fill in gaps of his memories with interviews with shipmates who might have been with him when he collected.[28] Darwin was sufficiently humbled by his lack of foresight.

> It never occurred to me, that the productions of islands only a few miles apart, and placed under the same physical conditions, would be dissimilar. I therefore did not attempt to make a series of specimens from the separate islands. It is the fate of every voyager, when he has just discovered what object in any place is more particularly worthy of his attention, to be hurried from it.[29]

Although Darwin's finches were of dubious importance to Darwin's ideas, the mockingbirds of the Galapagos were likely a great influence. He recognized that there were different species of mockingbird on the large islands of the Galapagos, with these species also occurring on the smaller islands around the large ones. Eventually, a fourth species would be discovered. Darwin immediately recognized the significance of these birds, and it was clear to him that the amount of variation of mockingbirds on the islands was greater than what he had observed in South America, and that the similarity of the mockingbirds in the Galapagos suggested a common ancestor that had come from the mainland. He wrote:

> These birds agree in general plumage, structure, and habits; so that the different species replace each other in the economy of the differ-

> ent islands. These species are not characterized by the markings on the plumage alone, but likewise by the size and form of the bill, and other differences.[30]

He was noting local adaptation not just in color but also in form. To show the use of his collections even today, the Museum of Natural History, London, produced a film on the specimens.[31] Although mockingbirds were common when Charles visited the Galapagos, the species on Isla Floreana was extirpated by 1880.[32] Mockingbirds were present on islands around Floreana, but to determine if they were the same species, tissue samples were taken from Charles's specimens for genetic analysis, and compared with the DNA of the other mockingbirds. Certainly Charles could not have known that his specimens would ever make such an important contribution because he knew nothing of DNA. This story suggests the importance of collections: we never know what the specimens we collect and deposit in museums will be used for in the future, and sometimes we can't even hazard a guess.

Back in England

Early in the voyage (June 1833), Darwin wrote, "It is a grand spectacle to see all nature thus raging; but Heaven knows every one in the *Beagle* has seen enough in this one summer to last them their natural lives."[33] Although Darwin would continue to make contributions to collections, his importance as a collector plummeted when he returned to the family home, "The Mount," on October 5, 1836. He was unannounced and his father and sisters were just sitting down to breakfast. Robert would eventually write, "Why the shape of his head is quite altered."[34] The person that Robert had chastened as someone who would bring shame to his family had matured and found his direction.

The idea of marriage soon came to Darwin's mind, and he approached finding a wife like collecting any other specimen. He wrote to his friend C. T. Whitley, May 8, 1838, as he was working on his contribution to *Narrative of the Surveying Voyages of His Majesty's Ships* Adventure *and* Beagle and editing *The Zoology of the Voyage of the H.M.S.* Beagle:[35]

> Of the future I know nothing I never look further ahead than two or three Chapters—for my life is now measured by volume, chapters & sheets & has little to do with the sun— As for a wife, that most interesting specimen in the whole series of vertebrate animals, Providence only know whether I shall ever capture one or be able to feed her if caught.

Infamously, Darwin would scribble in his notebook the pros and cons of marriage. Under "Marry," he wrote "constant companion, (friend in old age) . . . better than a dog anyhow."[36] He concluded "Marry, Marry, Marry." He married his first cousin, Emma Wedgwood, on January 29, 1839.

Charles had returned to England with some degree of fame and notoriety because Henslow had sent specimens to experts for identification and to describe new species. Darwin edited *The Zoology of the Voyage of the H.M.S.* Beagle, with five parts published between 1838 and 1843:

- I *Fossil Mammalia* by R. Owen, 1838–40 (preface and geological introduction by Darwin)
- II *Mammalia* by G. R. Waterhouse, 1838–39 (geographical introduction and a notice of their habits and ranges by Darwin)
- III *Birds* by J. Gould [and G. R. Gray], 1838–41
- IV *Fish* by L. Jenyns, 1840–42
- V *Reptiles* [and Amphibia] by T. Bell, 1842–43

The specimens obtained by Darwin were remarkable, and included a species named *Rhea darwinii* by Gould (Darwin's or lesser rhea, later synonymized with *Rhea pennata* and placed in the genus Pterocnemia). The holotype of *R. darwinii* has an interesting story. Darwin had heard of this smaller species of rhea but had been unsuccessful in finding a specimen. Early in January 1833, the *Beagle*'s artist, Conrad Martens, shot a rhea that they thought was a juvenile and they proceeded to eat it. Darwin then realized that it was not a juvenile but a specimen of the elusive smaller rhea. He managed to preserve the head, neck, legs, one wing, and some large feathers. The resulting holotype of *Rhea darwinii* was a near complete composite of several specimens.[37]

Darwin also collected plants, and perhaps it was plants that led Darwin to one of his most important colleagues, John D. Hooker, who would become the director of the Royal Botanical Gardens, Kew. Hooker read proofs of Darwin's *Beagle* accounts while on a similar voyage (1839–43), and was approached by Darwin upon his return to describe the Galapagos plant specimens. It was Hooker who probably first heard Darwin's evolutionary ideas, and Hooker who pushed Darwin to work on his most significant taxonomic project, a project that would shape the future of taxonomy and provide a forum for Darwin's developing evolutionary ideas. Charles wrote Hooker September 8, 1844, and told him of the ideas that he has been forming to explain the patterns of distributions that he had observed of species:[38]

> The conclusion, which I have come at is, that those areas, in which species are most numerous, have oftenest been divided & isolated from other areas, united & again divided;—a process implying antiquity & some changes in the external conditions. This will justly sound very hypothetical.
>
> I cannot give my reasons in detail: but the most general conclusion, which the geographical distribution of all organic beings, appears to me to indicate, is that isolation is the chief concomitant or cause of the appearance of new forms (I well know there are some staring exceptions) . . .

What Darwin is stating is that vicariance (geographic isolation) leads to speciation. This is now accepted as the main means of speciation, and it would have been considered heretical at the time. Hooker was skeptical, replying, "Nothing will give me so much pleasure as to get grounds for your reasonings & to carry out your theory of isolation."[39] Darwin would pass ideas off to Hooker and get most of his botanical information from him. Hooker, along with Thomas Henry Huxley, would come to the defense of Darwin at the great Oxford Evolution Debate, but before that, Hooker is responsible for sending Darwin down his eight-year foray into the taxonomy of barnacles.

By 1846, Darwin has cataloged all of his specimens except one, a barnacle. He wrote to Hooker, October 2, 1846, "Are you a good hand at inventing names: I have a quite new & curious genus of barnacle, which I want to name, & how to invent a name completely puzzles me."[40] It was an innocent beginning. Hooker's basic response was that he needed to complete the description of this species because Charles could not really talk about his ideas on evolution without doing some taxonomy. Hooker did not expect this to take eight years, and neither did Darwin, but taxonomy projects are like that. Once people heard that Darwin was working on barnacles, museums and collectors began sending him their specimens, and the project, like all good taxonomy, snowballed. In the end, Darwin published two monographs on the *Cirripedia*.[41] The taxonomy was much more akin to what taxonomists do today than anything that came before, and it hearkened back to the love of invertebrates and homology that Robert Grant had inspired in him at the University of Edinburgh. Darwin sought homologies to define his species, and the taxonomy was built around his idea of descent with modification.

CONCLUSION

Darwin now had all of the information he would need and all of the techniques that he required to finish *Origin of Species*. *Origin* is a book built on collections, and another collector, Alfred Russel Wallace, who had developed similar ideas as he had been working essentially as a professional collector, spurred him into publication. Darwin learned what he needed to know based on his experiences in the field and with specimens that he and others had collected. I, at least, see *Origin* as a culmination of the old school of taxonomy and collection's science. Darwin transformed taxonomy with his publication as he did all other fields of biology. Many fields of biology and geology can claim Charles, but more so than any other field of biology, we taxonomists can claim him as one of our own.

Darwin's work also illustrates the essential importance of fieldwork and collections-based science. Without his experiences in the field, and without careful examination of specimens, Darwin would not have been able to see the truths of the universe that he unveiled in *Origin*. Today, collections-based science is experiencing a Renaissance and perhaps its death sentence at the same time. Many universities are getting rid of their collections, and the collections are either being destroyed or being sent to large centers like the Smithsonian Institution, but it is the universities that are in charge of teaching the next generation of scientists. Darwin learned his love of his science from his professors, and if we continue to whittle away at the university collections, students studying natural history will also disappear. Large museums are important, but they cannot teach university students what they need to know.

Luckily, Alabama is far ahead of many regions. We still have excellent training of students in natural history, and the state supports two strong university museums at Auburn University and the University of Alabama. Perhaps it will be here that the next Darwin will be molded.

After Darwin's return from the voyage of the *Beagle*, he no longer was a significant contributor to collections; however, he did not stop. He continued to prepare specimens, particularly of his pets. Darwin had a fondness for fancy pigeons, with their strange morphology taking up a large chunk of Origin. As those birds aged and died, Darwin would skin them or skeletonize them, and even his pets now live on. Darwin lives on in his collections. Authors will cite the species he described for centuries, so he would have achieved some degree of immortality even if he had just been a taxonomist. Perhaps we would not be dedicating symposia to him or producing Hollywood movies about his life, but his work would still live on.

On Being Darwin

In the fall of 2008, James Bradley at Auburn University asked me if I would be interested in dressing as Darwin for the celebratory events in 2009. I have a passing resemblance with Charles, with a distinct brow ridge and what is left of my hair a sandy brown, but at 5'5" I was much shorter, and my weight tended more towards Robert than Charles, who had a relatively athletic build. I thought it would be fun, and I agreed, but I didn't want to be the long-white-haired, grizzled Darwin. Being only 39, I didn't think I could pull it off. Besides, it was a celebration of Darwin's birth and the publication of *Origin*, and so the Auburn University Theater

Department developed a period-accurate costume of Darwin from the time of the publication of *Origin*.

As I sat and watched the talks dressed as Darwin, I sometimes felt myself being more than an actor, and it seemed that when speakers talked of Darwin's discoveries, they were talking about me. I have come to realize that all biologists are like this. Darwin revolutionized biology, and nearly every biologist sees something in themselves of Darwin. It is easy to put on the costume and become one with the role.

Recently, having lost a lot of weight and being more in line with Charles's build, and needing to return finally my outfit, I decided to take Charles out for a walk. It was Halloween in downtown Auburn, so my appearance wouldn't be that unusual. During that walk, one woman asked if I was Samuel Adams. I told her I was Darwin, and she had no idea who Darwin was. I said "evolution," and she still had no idea. Exasperated, I said, "He was a biologist." What it taught me was that we still have a lot to do in teaching evolution, and there are a lot of people who still don't know what the science is. As we look at the debate that evolution still produces, we have to realize that the debate exists because education of the public is incomplete. ❧

NOTES

1 Cock, R. 2005. Scientific Servicemen in the Royal Navy and the Professionalisation of Science 1816–1855. In *Science And Beliefs*: *From Natural Philosophy To Natural Science, 1700–1900*. eds. D. Knight and M. Eddy. Aldershot: Ashgate, 95–111; Steinheimer, F. 2004. "Charles Darwin's Bird Collection and Ornithological Knowledge during the Voyage of H.M.S. Beagle, 1831–1836." *Journal of Ornithology* 145: 300–20.

2 Zirkle, C. 1941. "Natural Selection before the *Origin of Species*." *Proceedings of the American Philosophical Society* 84:71–123; Bayrakdar, M. 1983. "Al-Jahiz and the Rise of Biological Evolutionism." *Islamic Quarterly* 21:149–55.

3 Egerton, F. 2002. "A History of the Ecological Sciences; Arabic Language Science—Origins and Zoological." *Bulletin of the Ecological Society of America* 83: 142–46.

4 Bayrakdar, 1983. Quoted in Bayrakdar (1983).

5 Linnaeus, C. 1735. *Systema Naturae*. Leiden: Th. Haak.

6 Darwin, E. 1789. *The Botanic Garden, Part II, The Loves of the Plants*. London: J. Johnson; 1792. *Zoönomia; or, The Laws of Organic Life*. London: J. Johnson.

7 King-Hele, D. 1977. *Doctor of Revolution: The Life and Genius of Erasmus Darwin*. London: Faber & Faber.

8 Browne, J. 1996. *Voyaging*. Princeton: Princeton University Press.

9 Desmond, A. and J. Moore. 1991. *Darwin*. London: Penguin; Phipps, W. 2002. *Darwin's Religious Odyssey*. New York; Trinity Press International.

10 Desmond and Moore, 1991.

11 Phipps, 2002.

12 Burkhart, F. and S. Smith, eds. 1985. *The Correspondence of Charles Darwin, Volume 1, 1821–1836*. Cambridge; Cambridge University Press.

13 Darwin, C. 1887. *The Autobiography of Charles Darwin 1809–1882 Wih the Original Ommissions Restored. Edited and with Appendix and Notes by his Granddaughter Nora Barlow*. ed. N. Barlow. London: Collins, 1958.

14 Ibid.

15 Darwin, F., ed. 1887. *The Life and Letters of Charles Darwin, including an Autobiographical Chapter*, vol. 1. London: John Murray, 32.

16 Darwin, 1887.

17 Kohn, D., G. Murrell, J. Parker and M. Whitehorn. 2005. "How Henslow taught Darwin." *Nature* 436: 643–45.

18 Ibid.

19 Darwin, 1887.

20 Browne, 1996.

21 Ibid.

22 Darwin, C. 1839. *Narrative of the Surveying Voyages of His Majesty's Ships* Adventure *and* Beagle *Between the Years 1826 and 1836, Describing their Examination of the Southern Shores of South America, and the* Beagle*'s Circumnavigation of the Globe. Journal and Remarks, 1832–1836*. London: Henry Colburn, 11.

23 Darwin, 1839.

24 Owen, R. 1840. *The Zoology of the Voyage of H.M.S.* Beagle. *Part 1; Fossil Mammalia*. ed. C. Darwin. London: Henry Colburn.

25 Jenyns, 1842.

26 Darwin, C. 1845. *Journal of Researches into the Natural History and Geology of the Countries Visited During the Voyage of H.M.S.* Beagle *Round the World*, 2nd ed. London: John Murray, 388.

27 Darwin, 1845; Sulloway, F. 1982. "Darwin and his Finches: the Evolution of a Legend." *Journal of the History of Biology* 15:1–53; Steinheimer, F. 2004. "Charles Darwin's Bird Collection and Ornithological Knowledge during the Voyage of H.M.S. Beagle, 1831–1836." *Journal of Ornithology* 145: 300–20.

28 Sulloway, 1982.

29 Darwin, 1839.

30 Darwin, 1839.

31 Natural History Museum. Darwin's Mockingbirds Knock Finches off Perch. http://www.nhm.ac.uk/about-us/news/2008/november/darwins-mockingbirds-knock-finches-off-perch23090.html.

32 Curry, R. 1986. "Whatever Happened to the Floreana Mockingbird?" *Noticias*

de Galápagos 43:13–15; Grant, P. , R. Curry, and B. Grant. 2000. "A Remnant Population of the Floreana Mockingbird on Champion Island, Galápagos." *Biology Conservation* 92: 285–90.

33 Darwin, 1887.

34 Darwin, 1887.

35 Burkhart, F. and S. Smith, eds. 1986. *The Correspondence of Charles Darwin, vol 2: 1837–1843*. Cambridge: Cambridge University Press.

36 Darwin, 1887.

37 Steinheimer, 2004.

38 Burkhart, F. and S. Smith, eds. 1987. *The Correspondence of Charles Darwin, Volume 3, 1844–1846*. Cambridge: Cambridge University Press.

39 Ibid.

40 Ibid.

41 Darwin, 1851.

42 Darwin, 1854.

4

Darwin, Wallace, and Malthus

GERARD ELFSTROM

Charles Darwin (1809–82), Alfred Russel Wallace (1823–1913), and Thomas Malthus (1766–1834) had much in common. They were English, scientists, and lived in the 19th century (though Malthus's important work was completed late in the 18th century). Nonetheless, they led very different lives. Though his five-year voyage of discovery on the sailing ship HMS *Beagle* brought him fame and provided inspiration for his thinking on natural selection, Darwin suffered poor health for much of his life and remained close to his home near London. In contrast, Wallace spent many years trekking through exotic parts of the world tirelessly collecting specimens. Wallace devised a theory of natural selection independently of Darwin, but he was destined to remain in the shadow of Darwin's fame and distinction. Though nearly as famed as Darwin, but quite possibly more controversial, Malthus lived the quiet life of a college professor at the East India College in southeast England. In much the same way that the lives of these three men shared a great deal and differed a great deal, their ideas came together then diverged in ways that are fruitful and illuminating.

CONVERGENT THINKING

Everyone knows of Charles Darwin, the great scientist whose theories of evolution transformed biology and lit a firestorm of controversy that today continues burning vigorously. Many also know that his thinking was introduced to the world in his astonishing work, *On the Origin of Species by Means of Natural Selection,* which appeared in 1859. Fewer are aware that he began devising his revolutionary ideas more than 20 years earlier but was reluctant to publish them because he understood that they would prompt intense controversy. Even fewer are likely aware of the source of Darwin's inspiration. In his autobiography, Darwin described the crucial moment:

> In October 1838, that is, fifteen months after I had begun my systematic inquiry, I happened to read for amusement Malthus on Population, and being well prepared to appreciate the struggle for existence which everywhere goes on from long-continued observation of the habits of animals and plants, it at once struck me that under these circumstances favourable variations would tend to be preserved and unfavourable ones to be destroyed. The result of this would be the formation of new species. Here, then, I had at last got a theory by which to work.[1]

Malthus, in other words, supplied the threads that allowed Darwin to stitch his observations into a theory of natural selection. Though Darwin is commonly identified with the concept of evolution, that idea was widely understood in his day. Darwin's grandfather, Erasmus, was well aware of species change. But no one understood how the variation of species took place or why species varied. Darwin's theory of natural selection filled that void, and it remains his great achievement.[2]

Many readers also know of Alfred Russel Wallace. He was Darwin's younger contemporary, born in 1823, 14 years after Darwin. Even before *Origin of Species* appeared, Darwin was a towering figure in 19th-century science. The journals Darwin kept while voyaging on the *Beagle* were published before he returned to England, and they made his fame long before he again set foot ashore his native land. So, like many educated people, Wallace knew of Darwin and shared their high regard for his work. Though Darwin spent only a few years collecting specimens abroad, Wallace travelled for many years in remote and primitive locations, chasing down tens of thousands insects, birds, and plants.[3] In 1858 while recuperating from a bout of malaria on the island of Ternate, presently part of Indonesia, Wallace was struck by the following thought:

> One day something brought to my recollection Malthus's "Principles of Population," which I had read twelve years before. I thought of his clear exposition of "the positive checks to increase"—disease, accidents, war, and famine—which keep down the population of the savage races to so much lower an average than that of more civilized peoples. It then occurred to me that these causes or their equivalents are continually acting in the case of animals also; and as animals usually breed much more rapidly than does mankind, the destruction every year from these causes must be enormous in order to keep down the numbers of each species, since they evidently do not increase regularly from year to year, as otherwise the world would long ago have been densely crowded with those that breed most quickly. Vaguely thinking over the enormous and constant destruction which this implied, it occurred to me to ask the question, Why do some die and some live? And the answer was clearly, that on the whole the best fitted live. From the effects of disease the most healthy escaped; from enemies, the strongest, the swiftest, or the most cunning; from famine, the best hunters or those with the best digestion; and so on. Then it suddenly flashed upon me that this self-acting process would necessarily improve the race, because in every generation the inferior would inevitably be killed off and the superior would remain—that is, the fittest would survive.[4]

Excited by the implications of this insight, Wallace set to work on a paper in which he laid out his new theory of natural selection.[5] Completely unaware of Darwin's convergent thinking on this matter, Wallace sent him a copy for his examination. As is widely known, Wallace's article dissolved Darwin's fears and shocked him into rushing *Origin of Species* into print.[6]

Malthus's theory of population provided both Darwin and Wallace with the resource they needed to pull together their observations of species change and give them an explanation. It is entirely possible that one or both of them would have arrived at the principles of natural selection without Malthus's prompting. As it happened, Malthus provided each with the key he required.

Thomas Malthus died in 1834, barely 12 years after Wallace was born. Though trained as a minister, he spent most of his life teaching political economy at the East India College in Hertfordshire, England.[7] Kind and gracious in person, he was much beloved and highly respected by his contemporaries.[8] Malthus began developing his thoughts after an argument with his father on the question of whether human progress is possible. His father, an enthusiastic partisan of the Enlightenment, was confident of the possibility of human progress.[9] Like many sons, Thomas was keen to prove his father wrong. But, unlike many father-son arguments, Thomas spent the rest of his life working out the implications of his thinking and gathering evidence to support it. His theory of population initially appeared in 1798. As Malthus elaborated his views and accumulated data to support his theory, his early, brief work grew into a substantial book. It became one of the most widely read and influential works of the 19th century.[10] It appeared in many editions and can easily be found in print today. Like other educated people of their day, both Darwin and Wallace were well acquainted with Malthus's thinking.

The key Malthus provided to Darwin and Wallace is known informally as Malthus's "Dismal Theorem." Malthus described it early in *An Essay on the Principle of Population*. As he put it,

> I think I may fairly make two postula. First, that food is necessary to the existence of man. Secondly, that the passion between the sexes is necessary and will remain nearly in its present state. These two laws, ever since we have had any knowledge of mankind, appear to have been fixed laws of our nature, and, as we have not hitherto seen any alteration of them, we have no right to conclude that they will ever cease to be what they now are, without an immediate act of power in that Being who first arranged a system of the universe, and for the advantage of his creatures, still executes, according to fixed laws, all its various operations . . .
>
> Assuming my postulates as granted, I say, that the power of population is indefinitely greater than the power of the earth to produce subsistence for man. Population, when unchecked, increases in geometrical ratio.

> Substance increases only in arithmetical ratio. A slight acquaintance with numbers will shew the immensity of the first power in comparison to the second.[11]

As is apparent from the above passage, Malthus provided Darwin and Wallace with two essential ideas. One is that there is an upper limit to the population of any species. The second is the notion of a force, sexual passion, that propels the members of a species to reproduce until their numbers exceed that limit. Taken together, these two ideas gave Darwin and Wallace the intellectual framework they needed to construct the principle of natural selection.

A bit of reflection will make clear why Malthus's principle of population was considered "dismal," a term first applied to Malthus's work by the great British historian, Thomas Carlyle, in 1840.[12] Malthus's principle implies that it is absolutely impossible to create a human society in which each individual member can expect to enjoy a comfortable and productive life. In fact, a nearly perfect human society, that is, one able to completely eliminate suffering whether through violence or disease and to treat all citizens humanely, can nonetheless never escape the suffering which results from overpopulation in relation to food supplies.

Nearly a century after Malthus's work appeared, T. H. Huxley, known as "Darwin's Bulldog," made the point vividly. Huxley first imagined a utopian society as nearly perfect as human artifice can make it. As he formulated the matter in an 1893 lecture:

> Thus the administrator might look to the establishment of an earthly paradise, a true garden of Eden, in which all things should work together . . . But the Eden would have its serpent . . . Man shares with the rest of the living world the mighty instinct of reproduction, the tendency to multiply with great rapidity. The better the measures of the administrator achieved their object, the more completely the destructive agencies of the state of nature were defeated, the less would multiplication be checked . . . ?
>
> Thus, as soon as the colonists began to multiply, the administrator

> would have to face the tendency of the reintroduction of the cosmic struggle into his artificial fabric . . .[13]

Notice that Huxley fully understands the distressing consequence of Malthus's "Principle of Population": the more perfect the human society and the more comfortable its members, the quicker they will reproduce and the sooner will their carefully engineered social fabric unravel. Social harmony will inevitably be dissolved by a desperate struggle for existence brought about by a food supply that can never increase as quickly as population.

THE IMPLICATIONS OF MALTHUS'S THEORY

Though Malthus's *Principle of Population* brought the three men together, their thinking diverged when they considered its implications.

Malthus believed his theory demonstrates that efforts to aid the needy members of a society are both futile and counterproductive. An increase in food supplies will only prompt an increase in population, which will continue expanding until human reproductive appetites overwhelm food supplies and the population crashes. As a practical consequence, Malthus advocated abolishing the British poor laws, which were the public assistance programs of his day. Malthus formulated his ideas in this way:

> I have reflected much on the subject of the poor-laws and hope therefore that I shall be excused in venturing to suggest a mode of their gradual abolition . . . To this end, I should propose a regulation to be made, declaring that no child born from any marriage, taking place after the expiration of a year from the date of the law, and no illegitimate child born two years from the same date, should ever be entitled to parish assistance . . . If any man chose to marry without a prospect of being able to support a family, he should have perfect liberty do so. Though to marry, in this case, is, in my opinion clearly an immoral act, yet it is not one which society can justly take on itself to prevent or punish; because the punishment provided for it by the laws of nature falls directly and most severely upon the individual who commits the act.[14]

Malthus was convinced that only the prospect of miserable death by starvation for themselves and their children can suffice to curb the human sexual appetite. Notice also that Malthus did not assert that human beings necessarily desire to reproduce. Instead, their passions are strictly sexual. Of course, for most of human history, there has been a tolerably close link between sexual activity and reproduction.

Darwin, on the other hand, drew a very different inference from the pressures of selection. He says,

> The following proposition seems to me in a high degree probable—namely, that any animal whatever, endowed with well-marked social instincts, the parental and filial affections being here included, would inevitably acquire a moral sense or conscience, as soon as its intellectual powers had become as well, or nearly as well developed as in man.[15]

As Darwin saw it, beings living in groups must develop certain personal qualities if they are to maintain the integrity of their society. Such qualities include sympathy for others, cooperativeness, and a sense of fairness. Without these traits, a group cannot cohere. On the assumption that certain species acquire evolutionary advantages from communal living, selective pressures would introduce the above qualities into the organisms. But, Darwin believed, if such beings also acquire sufficient intelligence, both moral principle and moral practice will eventually develop from these social instincts.

Wallace, though he greatly admired Darwin, was unwilling to follow him to this conclusion. He agreed with Darwin's view that natural selection operating on social creatures would generate the personal qualities of sympathy, cooperation, and fairness needed to preserve social cohesion. Nonetheless, later in his life, and much to Darwin's dismay, Wallace became a spiritualist.[16] One result is that he became convinced that natural processes cannot account for a spiritual matter such as morality. He said,

> We thus find that the Darwinian theory, even when carried out to its extreme logical conclusion, not only does not oppose, but lends decided

> support to, a belief in the spiritual nature of man. It shows us how a man's body may have been developed from that of a lower animal form under the law of natural selection; but it also teaches us that we possess intellectual and moral faculties which could not have been so developed, but must have another origin; and for this origin we can only find an adequate cause in the unseen universe of Spirit.[17]

So who is correct? As it happens, Darwin had a simple and elegant response to Wallace's concerns: intelligent social beings with sophisticated powers of communication will have ample reason to be concerned about their reputations. This is because they will recognize that they are likely to need the assistance and cooperation of others. But they will also realize that support from their fellow group members is unlikely to be provided if they are not thought cooperative or helpful by the other group members. After a time, the pressures of natural selection are likely to transform this concern for reputation into the traits of character that constitute an individual others will wish to assist. In other words, the concern for reputation will be transformed into the prompting of morality.[18]

Darwin's simple argument remains fruitful. It is presently the cornerstone of 21st-century research on the processes by which standards of moral conduct may have emerged. The recent works of Mark Hauser and Richard Joyce are notable examples of research for which Darwin's insight remains relevant and fruitful.[19] In addition, many know of the work of Frans de Waal, the noted primatologist and psychologist at the Yerkes Primate Research Center in Atlanta. His studies of primate societies have yielded fascinating results concerning their expressions of sympathy, cooperation, and fairness.[20]

Wallace was well aware of Darwin's argument and devised a forceful response. The nub of his position is that Darwin did not offer rigorous proof of his views.[21] Darwin himself understood that his train of thought was speculative. Of course, Darwin's lack of conclusive proof is not evidence for Wallace's spiritualism. Strictly speaking, the issue remains undecided. However, the contemporary researchers mentioned above have found Darwin's ideas more useful in their investigations. So, in that sense, Darwin's view has prevailed.

But what about Darwin and Malthus?

Malthus's "Principle of Population" implies that in all times and under all circumstances population must necessarily increase far more quickly than food production. No matter how small the initial human population or how vast the untapped natural resources at their disposal, human groups will always find that their propensity to reproduce must surpass their ability to increase food production. If Malthus is correct, human beings are well advised to short-circuit whatever naturally evolved empathy they may have for others. In Malthus's view, seeking to help needy people only makes the inevitable population crash more painful. That is, aiding the poor by giving them food will allow them to reproduce in larger numbers. So when this increase in population surpasses the land's capacity to produce food, even more people will be alive to suffer the fate of starvation. In his view, the poor laws only made the difficulties of overpopulation more devastating.

Human population trends from Malthus's day to the present time reveal a major difficulty with his argument. In Malthus's day, around 1800, the world's human population was approximately one billion. Though the world's human population remained relatively constant from human prehistory to the period of the Industrial Revolution, it has increased markedly from 1800 to the 21st century. Presently, the world holds well over six billion human beings, and the United Nations expects the human population to increase to around nine billion by the middle of this century.[22] But there have as yet been no massive human die-offs, at least not as the result of population outstripping food supplies. There has been starvation aplenty, but it has resulted from war, civil turmoil, or natural disaster rather than population outstripping food supplies. In fact, one distinguished researcher, Amartya Sen, believes that the most common cause of starvation in the world is not lack of food. Rather, he believes the problem is that people do not have money to buy it. So he proposes giving needy people money rather than food.[23]

Two additional factors add insult to Malthus's injury. First the number of the world's people living in extreme poverty, understood as living on less than $1 per day, has plummeted in the past 25 years. In 1981, approximately 1.5 billion of the world's people lived on $1 per day or less. By 2002, that number had been cut by 400 million. Most of the improvement has occurred in east and south Asia. A good bit of the improvement for people in those areas is the result of the spectacular economic progress of China and India in recent decades. Chen and Ravallion note that should this trend continue the number of people living on $1 or less per day will be cut in half yet again by 2015,[24] again a result of economic progress in east and south Asia. It is also noteworthy that in 1980 the world's human population was approximately 4.5 billion, but it has since grown to well over 6 billion.[25] So even though the world's human population increased by more than one-third during the last 30 years, the portion of the world living in extreme poverty decreased significantly. To be sure, a considerable number of people still live on between $1 and $2 a day.[26] Nonetheless, there is clear improvement despite the considerable increase in population.

The second difficulty for Malthus's theory is that a considerable portion of the world's human beings have been steadily escaping poverty and becoming members of the middle class. Conceptions of "middle class" differ. One researcher defines it as the circumstance in which individuals have at least one-third of their income remaining after paying for food and shelter. The extra one-third of income can then be employed for a variety of things. Many will dispose of it on consumer goods, such as refrigerators or air conditioners. However, many will also use it for health care or education, both of which can considerably increase their standard of living. Using this definition, researchers conclude that around half of the human beings in the world at the present time are members of the middle class.[27] Further, many are confident the human middle class will continue to increase as a portion of the human population.

NEO-MALTHUSIANISM

Despite these observations, Malthus's theory retains considerable appeal and is prone to enjoy resurgence in periods of human turmoil.[28] In particular, it springs back to life in times of economic distress. Hence, neo-Malthusianism reappeared in the turbulent 1960s, then again in the 1970s.[29] In 2008, a newspaper headline—"Malthus Redux: Is Doomsday

Upon Us, Again?"—demonstrated that neo-Malthusianism revived yet again during the global economic crisis of the first decade of the 21st century.[30]

Nonetheless, to salvage Malthus's theory, the neo-Malthusians introduced an important modification. Rather than claiming that population growth must *at all times and in all places* outstrip increases in food supply, the neo-Malthusians borrowed an idea from ecology, that of carrying capacity. For the neo-Malthusians, there must be a limit to the amount of food the earth can produce and, therefore, a limit to the number of human beings it can support. The clear implication is, if human population increases beyond the limit of maximum food production, a population crash must follow.[31] This claim must be right. The earth is not becoming larger. So land and water on the earth are certainly limited resources. If so, the amount of food the earth can produce must have an upper boundary. This reformulation of Malthus's position is clearly not weakened by the above data regarding population increase and declines in human poverty. Instead, to the extent that the earth's human population continues to grow, the limit must be drawing closer. In addition, the degradation of the environment by polluting air and water decreases the land available for food production, and thus brings the upper limit of food production closer still.

But this neo-Malthusian position prompts two questions. The first, obvious, question is: When will the limit of maximum food production be reached? After all, the fabled religious mystics who predict the end of the universe will also be correct someday. As the "Again" in the newspaper headline, "Malthus Redux: Is Doomsday Upon Us, Again?," indicates, Malthusian catastrophes have been declared a number of times in the past.[32] But in each case, humanity has been able to muddle through. Hence, the great difficulty is determining when the Malthusian catastrophe is likely to arrive. The second question is: should we live our lives now on the assumption that the boundary of maximum food production will be crossed soon?

Important though the above questions may be, there is clearly insufficient data to address the first question. Since the first remains beyond resolution, the second must remain in limbo as well. On the other hand, if we do not know when the limit of maximum food production will be crossed, and if the consequences of crossing it include enormous human suffering, we are prudent to begin the effort to limit human population growth at the present time. Of course, in addition to that prudent caution, there are additional benefits to the environment to be had by decreasing the strain on natural resources and reducing the burden of pollution additional humans will inevitably create.

REINING IN POPULATION GROWTH

However, if prudence dictates limiting population growth, the inevitable question is: What is the best way to halt the burgeoning increase in the number of human beings? There are several possibilities.

Malthus endorses the following approach. He says,

> In most countries, among the lower classes of people, there appears to be something like a standard of wretchedness, a point below which they will not continue to marry and propagate their species. This standard is different in different countries, and is formed by various occurring circumstances . . . In an attempt to better the condition of the labouring classes of society our object should be to raise this standard as high as possible.[33]

There is presently some data to support Malthus's approach. The Soviet Union collapsed in the early part of the 1990s and fractured into a number of separate states. Russia is the largest resulting nation. During this period, the Communist Party was ousted from power. In fact, Russia declared it illegal for a number of years. However, during its 70-odd years of rule, the Communist Party became so intricately intertwined with all facets of Soviet society that Russia's major institutions collapsed along with the party. Both governmental and economic chaos ensued. Given the stress of the era, the economic uncertainty, and the general demoralization of the population, it is no surprise that the Russian birthrate plummeted.[34] And it is worth noting on behalf of Malthus's notion of the subjectivity of misery, that, though they were indeed miserable, the inhabitants of the former Soviet Union enjoyed a far higher standard of living than people in many parts of the world.

Nonetheless, many have an understandable reluctance to make people miserable, particularly when many of the world's human beings appear to have ample misery. Aside from that concern, however, the subjectivity of Malthus's approach is troubling. As noted above, billions of people in the world are far poorer than were the Russians even at the worst of their economic and social chaos. Though they are quite likely distressed by their circumstances, ample data reveal that the poorest, most abject nations of the world also have the highest birthrates. Even if Russian society had not eventually recovered from the trauma of the early 1990s, it is entirely possible that Russians would eventually have become accustomed to their circumstances and returned to their usual childbearing practices.

As mentioned above, the nations of the world with the neediest populations commonly also have the highest birthrates. It is difficult to know what could be done to make people in those nations feel so miserable that their appetites for reproduction would shrivel. However, given their present state of deprivation, it is entirely possible such measures would necessarily be indefensibly cruel. So at the very least, it would be reasonable to determine whether any other options for population reduction remain.

However, modern technology has recast the Malthusian problem. The world now possesses effective and widely available methods of birth control. Malthus's assumption that sexual activity would inevitably result in reproduction is outdated. He presumed that the only way to short-circuit reproduction was to make people so miserable that they would lose interest in sexual activity. At present, of course, human beings are able to engage in sexual activity without necessarily risking reproduction. The passion between the sexes may well remain as strong as in Malthus's day, but there is no longer reason to believe sexual activity will inevitably result in more human beings. Even the demoralized Russians of the early 1990s employed birth control to prevent reproduction. That is, they *chose* not to have children.[35] If so, Malthus's question can usefully be recast as the question: What circumstances will prompt human beings to decide against reproducing?

One obvious way to attempt to convince people to forgo reproduction is the age-old technique of coercion. Concerned by its burgeoning population, the government of mainland China employed this approach with its one-child-per-family policy introduced in 1979. As the name implies, the government stipulated that each married couple could have one child only. This policy relied on both incentives and punishments to gain compliance by the Chinese people. Its menu of punishments included fines and property or job loss. For the most part, China's policy was enforced by local governments. Because there is no entrenched rule of law in China and government oversight was spotty, there was considerable variation in the execution of the one-child policy from district to district.

Nonetheless, the fertility rate in China prior to 1979 was 2.9 children per woman. Following the introduction of the one-child-per-family policy, the fertility rate sank to 1.7 children per woman, and remains at that level to this day. Though 1.7 is considerably more than 1 child per woman, it is entirely low enough to keep the Chinese population in check. Globally, the fertility rate needed to maintain a stable population is 2.1 children per woman. The 2 children replace the parents, while the .1 allows for the portion of children who do not survive long enough to produce children of their own. In fact, the Chinese government claims that its policy has resulted in 250 million to 300 million fewer births than would have occurred without its policy.[36]

The Chinese one-child policy has prompted a number of questions. First, some researchers wonder just how much of the reduction in fertility is due to the one-child policy and how much the result of other factors, such as urbanization or the emergence of a culture in which small families are desired.[37] Some authors note that the fertility rate in China had fallen considerably even before the one-child policy was put into effect. In addition, Deng Xiaoping, China's leader from 1978 to 1992, introduced wholesale economic reform during the same period. Large numbers of Chinese flocked from the countryside to the cities. Urban dwellers have lower birth rates (in part because they have less space to house a family) than those in rural areas. In addition, many young Chinese women left their villages to seek jobs in factories. They wished to earn sufficient money to allow them to return to their villages and begin families. As a result, some will remain unmarried during their prime childbearing years and some may not return to their villages.

A more pressing difficulty is that it is unlikely that the Chinese policy can be successfully exported. The policy requires significant resources of social control and entails far greater governmental intrusion into personal lives than is likely to be tolerated in many other parts of the world. So while the one-child policy may or may not be effective in China, it is unlikely that it could be successfully applied in other nations of the world with very

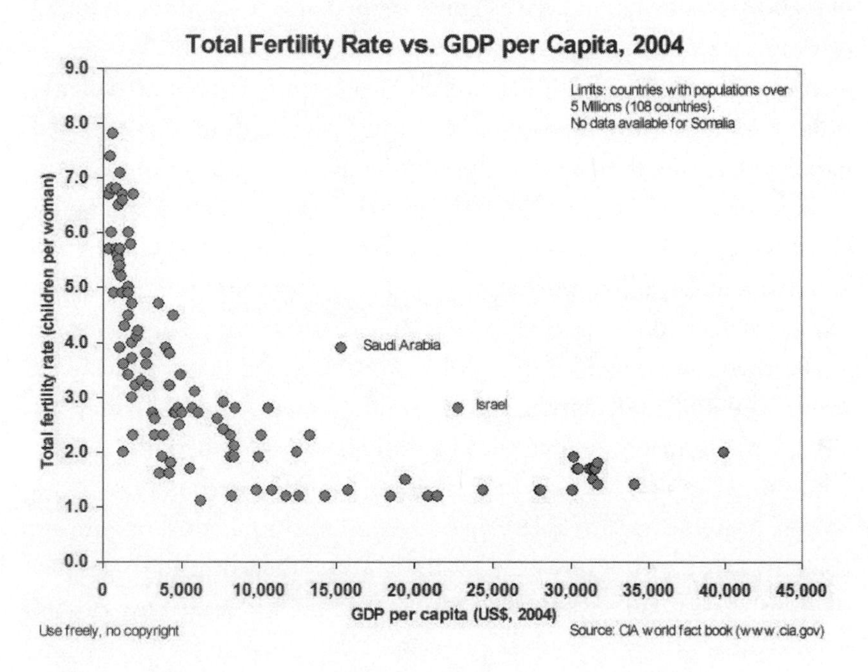

different cultures and very different systems of government.

Another approach to reducing the human fertility rate is suggested by this chart:

The chart strikingly illustrates an intriguing correlation. Notice that the chart's curve is concave. The fertility rate shoots well above 2.1 (the fertility rate necessary to maintain a stable population) when a nation's gross domestic product (or the total value of all the goods and services produced within the nation) drops below $5,000 per person. As the GDP increases above $5,000, the fertility rate drops well below 2.1.

That phenomenon is known as the demographic-economic paradox and is well documented. The term refers to the fact that fertility generally falls below the replacement level of 2.1 as national wealth increases. It is termed a paradox because many (including Malthus) assumed in the past that the more wealth people possess, the more children they will have. Nevertheless, the demographic-economic paradox has the happy implication that a most effective way to motivate human beings to forgo reproduction is simply to make people minimally financially solvent. A per capita GDP of $5,000 is quite modest, only slightly more than a tenth of the $46,859 of wealth per person enjoyed by the citizens of the United States.

But pleasing though this approach may be, the community of nations has not yet proven adept at increasing the wealth of impoverished nations. It is entirely true that the economies of India and China have made enormous strides in the past several decades. China's per capita GDP is now over $5,000, while India's has risen to slightly more than half that amount at $2,700. Yet no one knows how to transfer success to impoverished nations elsewhere.

Furthermore, it would be both helpful and illuminating to discover why a GDP of $5,000 apparently motivates human beings to have fewer children. Presumably, people with incomes at or above that level do not hold the belief that having many children is desirable. But there is no obvious reason why simply having a certain level of wealth should bring about this change. It is possible that the money brings other changes and that these other consequences are responsible for a new attitude toward childbearing.

As it happens, the example of the Indian state of Kerala holds a hint of how to address these questions. The citizens of Kerala are not even modestly wealthy. They have a per capita income of at most $810 per year, far below the $5,000 per year that triggers the demographic shift. Nonetheless, Kerala's citizens enjoy a life expectancy on par with that of the inhabitants of far wealthier nations. Other measures of the quality of life, such as average educational level and access to health care, are also extraordinarily high. And, perhaps not coincidentally, Kerala's fertility rate is approximately 1.8 per woman.[38] If Kerala is a good example, significant reductions in fertility can be achieved through measures that improve access to education and health care. If so, the key to the reduction of fertility is not money per se but the improvements in human life, such as education and health care, that higher incomes often allow.

A clue to Kerala's success is found in the United Nations Human Development Index. The first *United Nations Human Development Report* was issued in 1990. It contained a new measure of human development, the Human Development Index, which combined the elements of life expectancy, educational attainment, and income into a single, composite metric, the HDI. The HDI's achievement was the creation of a single statistic that can serve as a frame of reference to judge both social and economic development in the world's nations. The HDI sets a minimum and maximum for each dimension, called goalposts, and then shows where each country stands in relation to the goalposts, expressed as a value between 0 and 1.[39] At this point, there is likely little surprise that the nations ranking highest on the Human Development Index also have the lowest birthrates, while those ranking lowest on the HDI have the highest birthrates.[40] If so, it is entirely possible that Kerala's strong performance on the other measures of human development outweigh its very modest economic production.

Nonetheless, Amartya Sen has focused the picture still more tightly:

> There is considerable evidence that fertility rates tend to go down sharply with greater empowerment of women. This is not surprising, since the lives that are most battered by the frequent bearing and rearing of children are those of young women, and anything that enhances their decisional power and increases the attention that their interests receive tends, in general, to prevent over-frequent child bearing. For example, in a comparative study of the different districts within India, it has clearly emerged that women's education and women's employment are the two most important influences in reducing fertility rates. In that extensive study, female education and employment are the only variables that have a statistically significant impact in explaining variations in fertility rates across more than three hundred districts that make up India. In understanding inter-regional differences, for example, the fact that

> the state of Kerala in India has a fertility rate of only 1.7 (which can roughly be interpreted as 1.7 children on average per couple) in contrast with many areas which have four children per couple (or even more), the level of female education provides the most effective explanation.[41]

If Amartya Sen is correct, then the key factor in reducing human fertility is not increased wealth or even increasing human development as understood by the United Nations. Rather, the key is increasing educational and economic opportunities for women. So of the elements in the Human Development Index most pertinent to fertility reduction, providing education and economic opportunity to women appears to be the critically important factor. Apparently, as women become more educated and have more economic opportunity, they determine that their lives will be most satisfactory if they forgo having additional children. In happy consequence, a human community serious about reducing fertility rates may be able to do so without huge cost or complication simply by increasing the amount of schooling available to women and making it possible for them to participate in their economies.

CONCLUSION

Though sufficient misery may at times deter people from reproducing, the difficulty, as the subjectivity of Malthus's standard implies, is that people may become accustomed to any particular level of wretchedness. Furthermore, as the charts which map fertility against poverty indicate, the most miserable sectors of the world, such as Afghanistan, also have the highest fertility. The Chinese one-child program is possibly effective, but it is unlikely that it can be successfully exported.

So it seems the sociable qualities that Darwin believed are required for living in groups hold the key to preserving human life after all. If the observations of this chapter are correct, the same sympathy, cooperation, and sense of fairness that enabled ancient human beings to survive as social beings under conditions of great need may also help humanity survive its present-day challenges of prosperity and rapid human population growth. Simply increasing educational levels and economic opportunities for all human beings and working to ensure that they have adequate health care are key factors for reducing fertility. But, more precisely, increasing the educational level for women and opening doors to increased economic activity may be the centrally important factors. Not only can these things be easily provided by the world community, they are available at a decidedly modest cost. ⁂

NOTES

1 Barrett, P. and R. Freeman. 1989. *The Works of Charles Darwin*, vol. 29. London: William Pickering.
2 Lewontin, R. 2009. "Why Darwin?" *The New York Review of Books* 56 (9): 20.
3 Slotten, R. 2004. *The Heretic in Darwin's Court: The Life of Alfred Russel Wallace.* New York: Columbia University Press, 6.
4 Wallace, A. 1905. *My Life: A Record of Events and Opinions*, vol. 1. New York: Dodd, Mead and Company, 362.
5 Wallace, A. 1858. *On the Tendency of Varieties to Depart Indefinitely from the Original Type.* http://www.zoo.uib.no/classics/varieties.html.
6 Slotten, 2004.
7 Malthus, T. 1789. *An Essay on the Principle of Population*, 2nd ed. ed.P. Appleman. New York: W.W. Norton and Company, 2004.
8 Ibid.
9 Ibid.
10 Ibid.
11 Ibid.
12 Carlyle, T. 1840. *Chartism*. London: James Fraser, 109.
13 Huxley, T. and J. Huxley. 1947. *Touchstone for Ethics*. New York: Harper and Brothers, 50–1.
14 Malthus, 1789.
15 Darwin, C. 1871. *The Descent of Man*. Amherst: Prometheus Books, 1998, 101.
16 Slottin, 2004.
17 Wallace, A. 1889. *Darwinism: An Exposition of the Theory of Natural Selection With Some of Its Applications.* London: Macmillan and Company, 478.
18 Darwin, 1871.
19 Hauser, M. 2006. *Moral Minds: How Nature Designed Our Universal Sense of Right and Wrong*. New York: Ecco Press.
20 de Waal, F. 1996. *Good Natured: The Origins of Right and Wrong*. Cambridge: Harvard University Press. This work is a useful introduction to De Waal's research findings and thought.
21 Wallace, 1889.

22 United Nations Population Division. The World at Six Billion. http://www.un.org/esa/population/publications/sixbillion.html.

23 Sen, A. 1998. The Possibility of Social Choice. Lecture given for the Nobel Prize, Stockholm. http://nobelprize.org/nobel_prizes/economics/laureates/1998/sen-lecture.pdf.

24 Chen, S. and M. Ravallion. 2004. "How Have the World's Poorest Fared Since the Early 1980s? World Bank Policy Research Working Paper 3341." *The World Bank Research Observer* 19 (2). http://wbro.oxfordjournals.org/cgi/reprint/19/2/141.

25 United Nations Population Division, 1999.

26 Chen and Ravallion, 2004.

27 Parker, J. 2009. "Burgeoning Bourgeoisie." *Economist*, February 12. http://www.economist.com/specialreports/displayStory.cfm?story_id+13063298.

28 Malthus, 1789.

29 Hardin, G. 1968. "The Tragedy of the Commons." *Science* 162: 1243–48; Meadows, D., D. Meadows, et al. 1974. *The Limits to Growth.* New York: Universe Books.

30 McNeil, D., 2008. "Malthus Redux: Is Doomsday Upon Us, Again?" *New York Times*, June 15. http://www.nytimes.com/2008/06/15/weekinreview/15mcneil.html.

31 Hardin, 1968.

32 McNeil, 2008.

33 Malthus, 1789.

34 Perelli-Harris, B. 2006. "The Influence of Informal Work and Subjective Well-Being on Childbearing in Post-Soviet Russia." *Population and Development Review* 32 (4): 729–53.

35 Ibid.

36 Hesketh, T., L. Lu and Z. Wei Xing. 2005. "The Effect of China's One-Child Family Policy After 25 Years." *The New England Journal of Medicine* 353 (11): 1171–76.

37 Ibid.

38 Drèze, J. and A. Sen. 2002. *India: Development and Participation*. Oxford: Oxford University Press, 97–101; 112–42.

39 United Nations Development Program. Human Development Index. http://hdr.undp.org/en/humandev/hdi/.

40 Leete, R. and M. Schoch. 2003. "Population and Poverty: Satisfying Unmet Need as the Route to Sustainable Development." *Population and Poverty* 8: 9–37.

41 Sen, A. 2003. "The Importance of Basic Education." *Guardian*, October 28. http://www.guardian.co.uk/education/2003/oct/28/schools.uk4.

5

Sexual Selection

Debbie R. Folkerts

In his autobiography, Charles Darwin says of sexual selection that it was "a subject which had always greatly interested [him]," and that it was one of the few subjects which he had "been able to write about in full, so as to use all the materials which [he] had collected."[1] It is no wonder that he felt satisfaction with this idea because it makes such good, logical sense. In one way of thinking, it is easier to make sense of sexual selection than of most other forms of natural selection, since it deals directly with advantages in reproduction. Advantage in reproduction is what drives all of natural selection, Darwin's mechanism of evolution. Most of us think in terms of "survival of the fittest," where "fit" means having the means to find food, deal with the weather, and escape from predators. However, survival without reproduction has no effect on the genetic makeup of future generations. Survival is a means of ensuring the final goal of reproduction, and it is differential reproduction that makes the difference between fit and less fit types. The organism who leaves behind more offspring wins the contest.

Natural selection is the mechanism of adaptation, first described by Darwin[2] and Alfred Russel Wallace (1823–1913).[3] Driven by natural selection, populations change over time as certain types survive and reproduce more than other types and are therefore more successful at leaving behind offspring with their genetic characteristics. This can be simply stated as differential reproduction. Although Darwin and others have referred to sexual selection as an alternative to natural selection, the former is probably best understood as a special case of the latter, in which features that enhance an individual's mating success are favored, rather than or in addition to features that enhance survival. In other forms of natural selection, adaptations for survival are emphasized, but in sexual selection, adaptations for survival are often outweighed by characteristics involving mate acquisition. The principal idea of sexual selection was addressed in Darwin's early book, *On the Origin of Species*[4] but was covered more completely in his later book, *The Descent of Man, and Selection in Relation to Sex.*[5]

In Darwin's words, "sexual selection depends not on a struggle for existence in relation to other organic beings or to external conditions, but on a struggle between the individuals of one sex, generally the males, for the possession of the other sex."[6] For some time after he introduced the idea, sexual selection received more opposition than the idea of natural selection. Darwin was accused of inventing the idea because he was unable to explain the evolution of human characteristics through natural selection. Perhaps because of entanglement with the subject of human evolution or because of sex itself—both taboo topics—the idea of sexual selection has been somewhat controversial.

Ernst Mayr (1904–2005) suggested that much of the controversy and confusion has come from Darwin's sharp distinction between two kinds of selection, natural and sexual.[7] Should we consider the two to be separate phenomena? Darwin apparently did: "Sexual selection depends on the success of certain individuals over others of the same sex in relation to the propagation of the species; whilst natural selection depends on the success of both sexes, at all ages, in relation to the general conditions of life." There are, however, many cases where natural selection and sexual selection are both implicated and difficult to disentangle.

Mayr considered the distinction to depend on one's concept of fitness.[8] Many definitions have been proposed. Darwin's concept of "fit" means well-adapted for existence. Fitness for him had to do with survival of whole organisms, and he considered sexual selection a process that could lead to characteristics that are maladaptive (less fit) in a survival sense but that result in greater reproduction. Early mathematical geneticists redefined fitness in terms of single genes and their contributions to the gene pools of future generations whether from survival advantage or reproductive advantage.[9] Others refer to units of evolution and describe fitness as the "probability of leaving descendants."[10] Some understand fitness as a property of populations rather than of individuals.[11] It has been said that fitness includes several components, including viability (survival), fecundity (number of offspring), and fertility (probability of producing offspring). If sexual fitness is a component of overall fitness, then natural and sexual selection are not mutually exclusive, and the distinction is merely a point of view. Some have considered sexual selection to be minor compared to natural selection while others have demonstrated the power of sexual selection to mold population changes. The relative roles of survival fitness versus reproductive fitness depend on conditions of the environment in which selection is occurring. In most cases, both operate simultaneously. Although the argument may be mainly semantic, many modern scientists consider sexual selection to be a form of natural selection.

Understanding sexual selection requires a critical understanding of fitness. In sexual selection, we see fitness directly in terms of reproduction even when the characteristics seem to be counter-adaptive in terms of survival. Sexual selection explains the evolution of some of the most bizarre, puzzling, and paradoxical features of animals and plants, like the brilliant and gaudy plumage of some birds, the enormous antlers of some deer, and the luxuriously elaborate flower structures of many plants. Odd features for acquiring mates have evolved in many organisms even when those features impair survival by making escape from predators or food gathering more difficult or by requiring more food and energy to be spent in the process of acquiring a mate. Overgrown antlers may be nuisances to their possessors, but if they enable them to mate with more females and sire more offspring, the antlers are promoted by selection. Agonistic displays or actual fights between males require a considerable amount of energy and may result in injury or death but are promoted by selection because of advantages in mate acquisition. Sexual selection has even favored an odd copulatory suicide behavior in male redback spiders.[12] Males position themselves above the mouthparts of female spiders and comply with cannibalism because cannibalized males receive longer copulations, more fertilized eggs, and a greater likelihood that subsequent suitors will be rejected.

TYPES OF NATURAL SELECTION

Darwin described two types of sexual selection:

> The sexual struggle is of two kinds: in the one it is between the individuals of the same sex, generally the male sex, in order to drive away or kill their rivals, the females remaining passive; whilst in the other, the struggle is likewise between the individuals of the same sex, in order to excite or charm those of the opposite sex, generally the females, which no longer remain passive, but select the more agreeable partners.[13]

We now recognize these two forms of sexual selection as intrasexual selection, involving competition for mates between members of the same sex (usually males), and intersexual selection, involving a preference by one sex for individuals with certain characteristics in the opposite sex. Intrasexual selection has been more easily accepted and has received quite a bit of attention. Even Darwin's grandfather, Erasmus (1731–1802), recognized male-to-male competition.[14] On the other hand, the suggestion that sexual selection might occur through female choice was received with skepticism and even ridiculed by Wallace who thought the idea to be entirely speculative.[15] Many studies have now proven the existence of Darwin's female choice. The concept is well understood, although there is still much to discover about the nature of female choice and the criteria upon which females discriminate. Recent attention has been given to the idea of cryptic female choice, a phenomenon in mammals and birds in which a female can get rid of a male's sperm without his knowledge.[16] The equivalent in male-to-male competition is sperm competition, where even after mating, males may have differential reproductive success because of sperm number, size, speed, or subsequent matings.[17]

Even when there is obvious struggle among males or obvious choice exerted by females, there may be confusion as to whether sexual selection is occurring. Males often fight for territories rather than for possession of females; and females often choose males because of the quality of their territories. If both males and females choose quality territories that result in the well-being of their offspring, then natural selection rather than sexual selection is in operation.

Reproductive Strategies

Part of understanding these interesting phenomena lies in knowing that there is an inherent inequality between the sexes in terms of investment in reproduction and a resulting contrast in reproductive strategies. In most sexually reproducing organisms, there is a huge discrepancy in size of gametes. Eggs, by definition, are many times larger than sperm. Females, therefore, have a greater cost or investment in each egg than males do in each sperm. Females produce a limited number of eggs (which can be thought of as reproduction opportunities) at a relatively high cost for each one. Females may be even further limited in reproduction because of gestation and maternal care. Males, on the other hand, produce an almost unlimited number of sperm at a relatively low cost for each one and often engage in little or no parental care. This discrepancy has led to different strategies being typical of males and females. For males, a sensible strategy is to maximize reproduction opportunities, mate with as many females as possible, and produce as many offspring as possible, while investing relatively little in each one. For males, it is a contest of numbers. Females are more discriminating. A sensible female strategy is to make the most of the limited opportunities available by being selective about mates and therefore choosy in which genes to combine with hers. A female who makes a poor mating choice may have a total failure of reproduction and a total loss of her genes from future generations. Thus, there is strong selection pressure for females to be highly discriminatory in mate selection. Because a female will have few offspring relative to the potential for a male, she is also more likely to spend large amounts of time and energy ensuring the success of her offspring. This is why parental care, when it exists, is usually the responsibility of the female. Relative parental investment has been considered a key factor in sexual selection.[18] When one sex invests more, members of the other sex will compete among themselves to mate with members of the higher-investing sex. When investment is equal, sexual selection should operate similarly on both sexes.

There are interesting exceptions to the general rule of parental investment. In giant water bugs of the family *Belostomatidae*, for instance, females cement their fertilized eggs onto the backs of males. Males are, therefore, limited in reproduction by the number of eggs they can carry. Females are less limited because they may mate with a number of males. As a consequence, there is a reversal of parental roles as males are responsible for the care of developing eggs. Daddy water bugs fiercely protect their progeny from predators and aerate them at the water surface to prevent fungal growth. Another interesting example is found in sea spiders. Males possess specialized egg-carrying legs called ovigers. After mating, females entrust the care of their eggs to the well-equipped males and head off in search of other males to mate with. In these cases, the nature of the exceptions helps to prove the rule.

MATING SYSTEMS

Perhaps as a result of conflicting sexual strategies and varying environmental conditions, a number of different mating systems have evolved. Monogamy describes the mating of one male and one female. In such systems, parental care by both partners is typical and the pair usually mates for life. Both partners are equally limited in the number of offspring that are produced. A noted example is the Wandering Albatross, a large sea bird.

In polygamy there are multiple partners, either simultaneously or in a series. Polygyny, in which there are several females per male, as in a lion pride, is more common than polyandry, in which there are several males per female. Many species have mating systems with multiple females and multiple males, sometimes called promiscuity. Variation in environmental conditions is undoubtedly an important factor responsible for the variety of animal mating systems. For example, in environments with limited food resources, feeding of offspring by both parents, and therefore monogamy, may be most adaptive.

Julian Huxley (1887–1975) was the first to emphasize the important link between sexual selection and the nature of a species' mating system.[19]

He noted that exaggerated male morphology and display behavior are usually associated with polygyny. He also pointed out the importance of such environmental conditions as food resources and predation pressure in determining the adaptiveness of polygyny. Sexual selection is unlikely or even impossible in monogamous populations if the sex ratio is 1:1 and all individuals get paired. A likelihood of sexual selection exists, however, in populations that are polygamous and even in monogamous populations where a considerable proportion of individuals do not mate.

Sexual Dimorphism

In many species of animals, males and females are strikingly dissimilar in morphology. Males are often larger and stronger, more brightly colored, or possess adornments not possessed by females. When the sexes differ in features other than the character of their genitalia, the differences are considered to be secondary sexual characteristics. Primary sexual characteristics are those directly concerned with mating or gamete production. Darwin was interested in secondary sexual characteristics and noted,

> "The modifications acquired through sexual selection are often so strongly pronounced that the two sexes have frequently been ranked as distinct species or even as distinct genera. Such strongly marked differences must be in some manner highly important; and we know that they have been acquired in some instances at the cost not only of inconvenience but of exposure to actual danger."[20]

Much of sexual dimorphism can be explained by sexual selection, but even Darwin recognized that other forces may be involved in the evolution of differences between the sexes. A classical example is that of the extinct Huia, a bird that once lived in New Zealand. In that species, a striking difference in bill structure was the result of differential adaptation for feeding or, in Darwin's words, "differences in their habits of life." The Huia male used its thick bill to chisel away at wood for insects and spiders while the female's longer bill was able to probe deeper areas. Apparently, the bill difference allowed males and females to occupy different sub-niches and, therefore, reduce intersexual competition for food. We must recognize that sexual dimorphic features may be the result of a number of different selection pressures. Some are important in bringing the sexes together, some are related to competition (sexual or ecological), and some are related to sexual differences in rearing of offspring. Only when the differences result in reproductive advantage of certain individuals over others should they be considered to be evidence of sexual selection.

At least three categories of sexual dimorphism have evolved without sexual selection:

1. Characteristics that enhance mate finding or synchronize mating behavior and physiological readiness do not necessarily provide an advantage over other members of the same sex. The mustache markings of male Yellow-shafted Flickers, for example, are understood to enhance sex recognition. These characteristics evolve as a result of regular natural selection but may often be enhanced through sexual selection.

2. Characteristics that prevent mating with individuals of different species (reproductive isolation) are not typically the result of sexual selection. Hybrids usually have lowered fitness, and characteristics that prevent the formation of hybrids have been favored by natural selection. These include discrimination by females as well as any characteristics that make males more readily identified as members of the same species. The green head of mallards is understood in the context of species recognition. This point was almost missed by Darwin but was recognized by Wallace, who said such things enable "the sexes to recognize their kind and thus avoid the evils of infertile classes."[21]

3. Characteristics involved in ecological niche separation between the sexes, as previously mentioned, are not the result of sexual selection and were noted by Darwin: "When . . . the two sexes differ in structure in relation to habits of life, they have no doubt been modified through natural selection."[22]

Birds, Plants, and Primates

Studies of birds have been extremely important in the development of ideas about sexual selection. Darwin cited many bird examples of sexual selection and sexual dimorphism, in part perhaps because secondary sex characteristics may be more conspicuous in birds than in other animals.[23, 24] Birds were also important to Wallace.[25] Sexual selection explains the evolution of bright colors in males of species like the Northern Cardinal. Bright colors are often diet-related and present a relative cost to males in terms of food resources. The disadvantages of diet and of being more easily seen by predators are outweighed by the advantages of acquiring mates. Sexual selection also explains the even more bizarre adornments and behaviors seen in males of other bird species. In the peafowl, the brilliantly colored peacock displays his enormous tail feathers to the cryptically colored peahen, and bowerbird males spend inordinate amounts of time building and decorating bowers to attract females. Males in one group of bowerbirds build an "avenue" bower made of two walls of vertically placed sticks. The male collects hundreds of brightly colored objects—including shells, leaves, flowers, feathers, stones, berries, plastic items, coins, nails, rifle shells, or pieces of glass—and spends hours arranging his collection in and around the bower. Birds continue to inform ecological and ethological research, providing the basis for modern interpretations of sexual selection.[26] Studies of Eastern bluebirds being conducted at Auburn University are revealing the factors upon which female birds make their choices of mates and the benefits of reproduction yielded by their choices.[27]

Not until recently have sexual selection ideas been considered by botanists.[28, 29] Yet competition for mates is probably as important to the evolution of some higher plants as it is to animals. Botanists now refer to scrambles among pollen to reach stigmas and fertilize ovules. The results of male-to-male competition in plants include attraction and reward of pollinators through large showy flowers. Evidence also exists that in some plants female choice dictates which pollen tubes successfully grow down the style to reach and fertilize ovules. A new definition of sexual selection has been proposed in which sexual selection is simply viewed as selection that arises from differences in mating success.[30] This definition can be applied to all organisms and is consistent with Darwin's writings.

Of all the taxa that interested Darwin, the primates were most important for his ideas about human descent. He discussed the evolution of human secondary sexual characteristics within the context of mammal and, especially, primate evolution.

In primates, differences between the sexes often involve weight and muscular development, body dimensions, and pelage color, as well as such anatomical features as long canines, bulbous noses, gular pouches, shoulder capes, crests, baldness, and skin ridges.[31] Sexual selection in primates, as in most higher animals, is most evident in species with polygynous mating systems. The most marked sexual dimorphism of all primates occurs in the Old World monkeys, including many familiar species such as baboons and macaques. Many of their secondary sexual characteristics are associated with the highly competitive nature of their societies, in which members of both sexes frequently gesture submissively when dominant males display. Adaptations resulting from both environmental selection pressures and social conditions are obvious in the primates. Both have operated together throughout the history of primate evolution. Primate societies provide for cooperation and greater effectiveness in foraging for and protecting the young but also exert selection pressures of a different kind. In most present-day nonhuman primates, a few males produce most of the offspring. There is competition among males but little evidence of female choice.

Sexual Selection in Humans

Have humans been subject to similar kinds of selection pressures? It is almost certainly a fact that we have; however, the topic has been somewhat controversial and direct evidence is difficult to obtain. Darwin conjectured that the male beard, as well as the relative hairlessness of humans compared to nearly all other mammals, are results of sexual selection.[32] He also hypothesized that sexual selection could have caused the differentiation between human races, as he did not believe that natural selection provided a satisfactory answer. It has been hypothesized that many human behaviors not clearly tied to survival benefits—such as humor, music, visual art, verbal creativity, and some forms of altruism—are courtship adaptations that have been favored through sexual selection.[33] In fact, researchers have proposed that the large size of the human brain may not have been selected for in terms of survival and is probably a sexually selected trait.[34] Human vocabulary on average contains more words than necessary for communication and is believed to be used as a way to demonstrate intelligence and other forms of fitness to potential mates.[35] Even attractiveness of the human face has been investigated in terms of sexual selection and it has been said that a sense of human beauty results at least in part from symmetry because it is an indicator to potential mates of health, immunocompetence, developmental homeostasis, and parasite resistance.[36]

Some consider that our courtship behaviors have much in common with courtship of birds and other mammals. Sexually dimorphic characteristics in humans include size, strength, muscular development, metabolic rate, energy utilization, longevity, sexual maturity, sexual drive, complexion, hair pattern, breast development, and voice.[37] At least some of these correlate with those found in other primates. However, humans differ from other primates in the evolution of hair loss, increased skin sensitivity, a behavioral emphasis on contact comfort, loss of estrus, and female orgasm. These are said to relate to a specifically human grade of social organization.[38] The complex of human courtship and social interaction appears to be the result of long-term adaptation to environmental and social conditions in which man developed a polygamous and later a mainly monogamous mating system with effective rearing of young in social units of small size with a tendency for long-term pair bonding. Almost all of this evolution occurred prior to the emergence of complex civilizations. It is no wonder that we now struggle with the role of sexual differences in modern society.

Models of Sexual Selection

The quest to solve the mysteries of sexual selection seems endless, not because Darwin was wrong but because the details of how the process is driven in different systems continue to elude us. Perhaps also because we are fascinated by our own sexuality, scientists today continue to study the phenomenon of sexual selection in the context of modern understanding of genetics, ecology, behavior, and sociology.

Since Darwin first presented the idea, biologists have recognized the importance of sexual selection but have struggled to understand the precise mechanisms. The evolution by sexual selection of traits with a survival disadvantage has been considered a paradox. Many scientists have asked how the process gets started and how a trait becomes exaggerated. R. A. Fisher's (1890–1962) hypothesis of runaway selection has been popular to explain this puzzle.[39] His idea is that female preference for a certain trait may start arbitrarily, but once it has become established in a population, the trait and the preference will evolve together in an explosively positive feedback loop. The mechanism involves the fact that females with a strong preference for the particular trait, regardless of any survival handicaps the trait may infer, will have a reproductive advantage because their sons will possess the trait and, therefore, be attractive to other females with the same preference. In runaway selection, the strength of female preference increases over generations as females with a stronger preference fare better because of the reproductive success of their sons. The degree to which the trait is expressed in males also increases over generations as the trait becomes more and more highly preferred and may progress well beyond the point of survival fitness. Any heritable trait may be subject to this process because it is not the trait itself but the fact that females choose the trait that makes it attractive. Both the trait and the preference may continue to advance until at some point they may be counterbalanced by natural selection for survival fitness. An elaboration of this concept was coined the "sexy son hypothesis."[40] In particular, the sexy son hypothesis implies that any other benefits a potential mate may have to offer, compared to the sexual attractiveness of their sons, are irrelevant to the success of a female's offspring. Since a conceptual model of the runaway process was proposed, others scientists have modeled mathematical proof of the idea.[41, 42]

As a catchy response to the runaway model, a "chase-away model"[43] of sexual selection proposes that exaggerated male traits evolve as a consequence of an evolutionary arms race between males and females for control of reproduction.

Other hypotheses have been developed to explain sexually attractive traits with survival impairment in which females receive survival benefits for their offspring as a result of their choosiness. The "handicap hypothesis"suggests that a male's ability to survive in spite of a handicap indicates overall fitness and good genes to contribute to offspring.[44] The so-called "good genes model"also considers genetic benefits received by choosy females but still assumes a dichotomy between mating success and survival success.[45] The "truth-in-advertising model"proposes that sexual selection favors the evolution of costly and variable traits, whether they are handicaps or not, because their expression reflects the survivorship, vigor, and overall genetic quality of males who possess them.[46] Females choose males with exaggerated expression of preferred traits because they truthfully advertise desirable attributes that can be passed on to offspring. Others have emphasized the role of parasites in sexual selection by suggesting that females prefer to mate with males with showy sexual displays because they are the healthiest and the most resistant to parasites.[47] By avoiding infected males, choosy females can reduce their risk of contracting contagious diseases, obtain more parental investment, and increase the resistance of their progeny to parasites. The idea of "fluctuating asymmetry"is based on observations that healthier specimens have more left-to-right-sided symmetry than less healthy specimens.[48] Studies with rodents and humans now suggest that chemosensory signals in an individual's scent reveal a male's disease resistance and genetic compatibility.[49] Recently an approach has been encouraged that incorporates multiple sexual selection mechanisms, considers a continuum of reproductive and survival fitness, and exploits advances in physiology and molecular biology to understand the mechanisms through which both males and females achieve reproductive success.[50, 51] None of these new models of sexual selection contradict the model as originally proposed by Darwin. It is remarkable that 150 years since his first publication on the idea, modern science has just refined the basic idea.

In contemplation of his life's work and having listed each of his books, Darwin says in his autobiography, "I think that I have become a little more skillful in guessing right explanations and in devising experimental tests; but this may probably be the result of mere practice, and of a larger store of knowledge." This is a humble statement from a man so highly revered by biologists since his time. Many scientists have considered Darwin's idea of sexual selection to be one the most profound insights in all biology. This idea, possibly a favorite of Darwin's and still popular today, has allowed us to understand the evolution of some of biology's most interesting phenomena and continues to stimulate investigations into the complexities of life. ☙

NOTES

1 Darwin, C. 1887. *The Autobiography of Charles Darwin 1809–1882*. ed. N. Barlow. New York: Norton, 1958. This book was edited by Charles's granddaughter, Nora, with original omissions restored, and includes notes and letters.

2 Darwin, C. 1859. *On the Origin of Species By Means of Natural Selection, or the Preservation of Favoured Races in the Struggle for Life*. London: John Murray.

3 Alfred Russel Wallace independently conceived of an idea very similar to Darwin's natural selection and wrote a letter about it to Darwin. Darwin's friends arranged for a simultaneous announcement of their idea at the Linnaean Society in London in 1858. Wallace's 1858 essay was entitled, "On the tendency of varieties to depart indefinitely from the original type."

4 Darwin, 1859.

5 Darwin, C. 1871. *The Descent of Man, and Selection in Relation to Sex*. London: John Murray.

6 Ibid.

7 Mayr, E. 1972. Sexual Selection and Natural Selection. In *Sexual Selection and the Descent of Man: The Darwinian Pivot.* ed. B. Campbell. New Brunswick: Aldine Transactions.

8 Ibid.

9 J.B.S. Haldane, R.A. Fisher, and S. Wright were mathematical geneticists who published works in the early 1930s involving models of evolution.

10 Thoday, J. 1953. "Components of Fitness." *Symp. Soc. Exptl. Biol.* 7: 96–113.

11 Fisher, R. 1930. *The Genetical Theory of Natural Selection*. Oxford: Clarendon Press.

12 Andrade, M. 1996. "Sexual Selection for Male Sacrifice in the Australian Redback Spider." *Science* 271: 70–72.

13 Darwin, 1871.

14 Darwin, C. 1794. *Zoonomia*. London: J. Johnson.

15 Wallace, A. 1889. *Darwinism: An Exposition of the Theory of Natural Selection and Its Applications.* London: Macmillan.

16 Eberhard, W. 1996. *Female Control: Sexual Selection by Cryptic Female Choice.* Princeton: Princeton University Press.

17 Parker, G. 1970. "Sperm Competition and its Evolutionary Consequences in the Insects." *Biological Reviews* 45: 525–67.

18 Trivers, R. 1972. Parental Investment and Sexual Selection. In *Sexual Selection and the Descent of Man: The Darwinian Pivot.* ed. B. Campbell. New Brunswick: Aldine Transaction.

19 Huxley, J. 1938. "Darwin's Theory of Sexual Selection and the Data Subsumed by It." *The Light of Recent Research. American Naturalist* 72 (742): 416–33.

20 Darwin, 1871.

21 Wallace, 1889.

22 Darwin, 1871.

23 Ibid.

24 Selander, R. 1972. Sexual Selection and Dimorphism in Birds. In *Sexual Selection and the Descent of Man: The Darwinian Pivot.* ed. B. Campbell. New Brunswick: Aldine Transaction.

25 Wallace, 1889.

26 Selander, 1972.

27 Liu, M., L. Sieferman, H. Mays Jr., J. Steffen and G. Hill. 2009. "A Field Test of Female Mate Preference for Male Plumage Coloration in Eastern Bluebirds." *Animal Behaviour* 78 (4): 879–85. Geoff Hill and students at Auburn University have studied Eastern bluebirds for several years. Much of their research deals with sexual selection.

28 Willson, M. 1994. "Sexual Selection in Plants: Perspective and Overview." *American Naturalist* 144: S13–S39.

29 Andersson, M. and Y. Iwasa. 1966. "Sexual Selection." *Trends in Ecol. & Evol.* 11 (2): 53–8.

30 Arnold, S. 1994. "Is There a Unifying Concept of Sexual Selection that applies to both Plants and Animals?" *American Naturalist* 144: S1–S12.

31 Crook, J. 1972. Sexual Selection, Dimorphism, and Social Organization in the Primates. In *Sexual Selection and the Descent of Man: The Darwinian Pivot.* ed. B. Campbell. New Brunswick: Aldine Transaction.

32 Darwin, 1871.

33 Miller, G. 2000. *The Mating Mind: How Sexual Choice Shaped the Evolution of Human Nature.* London: Heinemann.

34 Schillaci, M. 2006. "Sexual Selection and the Evolution of Brain Size in Primates." *PloS One* 1(1). http://www.ncbi.nlm.nih.gov/pmc/articles/PMC1762360.

35 Miller, 2000.

36 Grammer, K. and R. Thornhill. 1994. "Human (Homo sapiens) Facial Attractiveness and Sexual Selection: The Role of Symmetry and Averageness." *Jour. Comp. Psychology* 108 (3): 233–42.

37 Caspari, E. 1972. Sexual Selection in Human Evolution. In *Sexual Selection and the Descent of Man: The Darwinian Pivot.* ed. B. Campbell. New Brunswick: Aldine Transaction.

38 Ibid.

39 Fisher, 1930.

40 Weatherhead P. and Robertson R. 1979. "Offspring Quality and the Polygyny Threshold: 'The Sexy Son Hypothesis.'" *American Naturalist* 113 (2): 201–08.

41 Lande, R. 1981. "Models of Speciation by Sexual Selection on Polygenic Traits." *PNAS* 78: 3721–25.

42 Kirkpatrick, M. 1982. "Sexual Selection and the Evolution of Female Choice." *Evolution* 36 (1): 1–12.

43 Holland, B. and W. Rice. 1998. "Perspective: Chase-Away Sexual Selection: Antagonistic Seduction versus Resistance." *Evolution* 52 (1): 1–7.

44 Zahavi, A. 1975. "Mate Selection –A Selection for a Handicap." *Journal of Theoretical Biology* 53: 205–14.

45 Hamilton, W., and M. Zuk. 1982. "Heritable True Fitness and Bright Birds: a role for Parasites?" *Science* 218, 384–87

46 Kodric-Brown, A. and J. Brown. 1984. "Truth in Advertising: The Kinds of Traits Favored by Sexual Selection." *American Naturalist* 124 (3): 309–23.

47 Hamilton and Zuk, 1982.

48 Moller, A. and A. Pomiankowski. 1993. "Fluctuating Asymmetry and Sexual Selection." *Genetica* 89 (1–3): 267–79.

49 Penn, D. and W. Potts. 1998. "Chemical Signals and Parasite-Mediated Sexual Selection." *Trends in Ecol. & Evol.* 13 (10): 391–96.

50 Kokko, H., R. Brooks, J. McNamara, and A. Houston. 2002. "The Sexual Selection Continuum." *Biological Sciences* 269: 1331–40.

51 Zeh, J. and D. Zeh. 2003. "Toward a new Sexual Selection Paradigm: Polyandry, Conflict and Incompatibility." *Ethology* 109: 929–50.

6

Human Evolution

Where Are We? Where Are We Going?

Shawn Jacobsen

The year is 2070. Millions of microscopic robots swarm through Joe Imdton's bloodstream, crawling through his cell membranes, repairing DNA mutations, disposing of wastes, repairing proteins, extending telomeres. Joe has a computer the size of a raindrop attached to his brainstem that gives him wireless access to all the world's information and entertainment. Technology has removed disease, hunger, pain, and depression and has given, in return, happiness and near immortality.

But there is still emptiness. Joe longs for something more than his neurochip and its mood-enhancing software provide, something even beyond the tranquility and reverence he felt while kneeling as a boy in the blue and red lights of his church's stained glass windows. The still-organic part of him seeks and believes in a final and magnificent climax to the experience of being human.

So, for the sake of Joe and the rest of humanity, on this day the wisdom and consciousness of all but the few unwilling will unite through a gravity wave interface. Humanity will become one mind. Some even believe they will become as God—an omnipotent intelligence spreading through the universe as a spiritual presence and perhaps even adopting as their children intelligent life forms on other planets.

The entire planet awaited, assembled not in the square of Earth's capital or at the base of a sacred mountain, but as they had assembled for the last 50 years, as one people individually connected to their home computers.

ON MARCH 28, 1838, Charles Darwin, then 29 years old, traveled to the London Zoo to visit an orangutan named Jenny. The morphological similarities he observed between Jenny and humans marked the beginning of decades of Darwin's musing on the common ancestry of humans and apes. Over the years, Darwin kept his ideas on human evolution private for fear the concept would reinforce prejudices against the general evolutionary theory he presented in 1859 in *Origin of Species*. But gradually his thoughts developed more structure until, in 1871, he completed and published *The Descent of Man, and Selection in Relation to Sex*. In this volume, he presented, in his words, "The Evidence of the Descent of Man from some Lower Form." This he based, firstly, on his three widely known criteria:

1. There is variation in human characteristics.
2. These characteristics can be passed on to offspring.
3. Some variations are more likely to provide their owners with a reproductive advantage than others.

He also presented "three great classes of facts" as evidence for human evolution:

a. similarity in bodily structure between man and other mammals,

b. similarity in embryonic development between man and other mammals, and

c. rudimentary organs—organs inherited from ancestors and of limited or no function.

With *Origin of Species* and *The Descent of Man*, Darwin addressed the question of "where did we come from," and in answer opened for the world the new science of evolution.

Have we evolved since becoming modern *Homo sapiens*? Are we still evolving? What about the future? What about Joe? Some futurists believe his circumstances as imagined above in 2070 are realistic, even probable. If that's true, will human evolution end before humanity does?

THEN

By 50,000 years ago, the brains of our ancestors had swelled to their current size of 1,300 cubic centimeters, and their jaws had become smaller and weaker. They had lost most of their hair, and their children were ridiculously helpless. Their larynxes had sunk so low in their vocal tracts that many choked to death, but they had gained the ability to produce a greater variety of sounds than any other primate. They were physically, and probably intellectually, fully human.[1] They made tools and used fire. By 40,000 years ago they drew paintings on cave walls, and by 20,000 BC they had populated every major land mass except Antarctica.

Some authorities claim that human evolution came to a virtual standstill by the time people were creating art, about 40,000 years ago. Still others believe our evolution ended when modern medicine and sanitation interfered with the harsh selection processes our species had grown up with.

There is little fossil evidence to support that human evolution has continued over the last 50,000 years. Over this geologically short time span, we must view evolution in terms of changes in gene frequencies. Any increase or decrease in the frequency of a gene (for example, one that codes for resistance to smallpox) in a population constitutes evolution. Genetic drift will cause gene frequencies of any population to change from generation to generation by random chance. Of greater interest is this question: have we recently felt the forces of natural selection? That is, have we changed, not by random chance, but to adapt to a changing environment?

Some believe the answer is, if so, probably little. However, some scientists, such as John Hawks of the University of Wisconsin at Madison, believe that the popular view of our "static" evolution is not only wrong but that recent evolution has possibly sped up. University of Utah anthropologist Henry Harpending believes evolution continues to mold us right up to the present. "We aren't the same as people even a thousand or two thousand years ago," he says.[2] These two scientists estimate that over the past 10,000 years humans have evolved 100 times faster than in any time in the past six million years—when we diverged from our closest living relative, the chimpanzee. They attribute this acceleration to change in diet, new diseases, and poor sanitation associated with civilization.[3]

Geneticists Robert Moyzis and Eric Wang of the University of California at Irvine believe there was a lot of evolving going on as humans expanded across the varied environments of Australia, Europe, Asia, and the Americas between 80,000 and 20,000 years ago. Using new methods

Highlights of Human Evolution

Years Before Now	Highlight
5–7 million	Divergence of human and chimpanzee lineages
4.4 million	Ardipithecus ramidus (350 cm^3 cranial capacity, first bipedal primate)
4 million	First Australopithecus (450 cm^3 cranial capacity)
2.3–2.6 million	Oldest stone tools
1.8 million	First Homo erectus (average 1000 cm^3 cranial capacity, controlled fire)
1.5 million	Homo erectus had left Africa and populated parts of Asia
400,000	Homo erectus and Homo neanderthalensis split
200,000	First Homo sapiens (eventually to 1300 cm^3 cranial capacity)
100,000	Homo sapiens began to leave Africa eventually replacing all existing Homo.
70,000	Development of genes associated with speech (although complex language is as old as 300,000 years)
35,000	Art, possible religious beliefs
30,000	Neanderthals extinct
20,000	Homo sapiens have populated every major land mass except Antarctica.
11,000	Beginning of agriculture and domestication of animals
5,800	Beginning of civilization
➡	
Year 2070	Fusion of human consciousness with intelligent machines and beginning of saturation of universe with human intelligence

FIGURE 6.1

Highlights of human evolution. There are many more species of hominids (members of the family *Hominidea* which evolved after our split with chimpanzees) including *Orrorin, Sahelanthropus, Kenyanthropus, Paranthropus*, two species of *Ardipithecus*, several species of *Australopithecus*, and several species of *Homo*. At various points in history, there were as many as six or seven hominids living at the same time. *Australopithecus* and *Homo erectus* are specifically mentioned in this table because they are the most likely to be our direct ancestors.

for analyzing genetic data, they conclude that 7 percent of human genes fit the profile for recent change, most of these within the last 40,000 years. They claim the rate of these changes has increased exponentially along with the world's population.[4]

Hawks, involved in the same project as Moyzis and Wang, agrees. He says genomes of people in this study had a high number of unique genetic traits associated with adaptations such as resistance to cold and infectious diseases. He and his collaborators concluded that by 5,000 years ago, humans were evolving 30 to 40 times faster than they ever had.[5]

Consider the following partial list of recent human adaptations: The advent of light skin within the last 20,000 years allowed humans to better manufacture vitamin D in sun-deprived climates. An 8,000-year-old mutation in Northern Europe allowed people to digest lactose and take advantage of the plentiful food source provided by dairy cows. Malaria is only 35,000 years old, but in Sub-Saharan Africa and other malaria regions, 25 new malaria resistant genes have been identified. The microcephalin gene, which affects brain size, arose 37,000 years ago, and ASPM, a gene also involved in regulating brain size, is only 5,800 years old, about the same age as civilization.[6]

Eduardo Ruiz-Pesini and coworkers discovered a mitochondrial DNA mutation that makes mitochondria (the cell's energy producers) less efficient so they create more heat and less ATP (energy).[7] The mutation is present in 75 percent of people in arctic regions but in only 14 percent of people from temperate regions. It was not present in any of the 1,000 study subjects from Africa. The selective advantage of this gene in cold climates is obvious. The researchers state that this mutation evolved within the last 50,000 years.

Researchers found that a group of Tibetan women with 10 percent higher oxygen-carrying capacity share a variation of a single responsible gene. They have, on average, more children than other women. The selective advantage of this gene is apparent as these people live in the thin atmosphere of 4,000 meters (13,000 feet) above sea level.[8]

Children with attention-deficit/hyperactivity disorder are more likely to have a particular variant of the gene DRD4. Sequencing studies show that this variant arose 50,000 years ago and is more prevalent farther along the ancestral human journey out of Africa. The gene is found in 80 percent of South Americans but in only 20 percent of Africans and Europeans. Children with ADHD score higher on tests for novelty-seeking and risk-taking, possible adaptations to exploring new frontiers.[3]

Now

In his book *Redesigning Humans*, Gregory Stock states,

> In the past, the reproductive isolation needed to generate even the modest biological differences among human groups has required geographical or cultural separation. Both, however, are greatly diminishing because of increased individual mobility, modern communications, and softening cultural rigidities . . . Traditional Darwinian evolution now produces almost no change in humans and has little prospect of doing so in the foreseeable future.[9]

Steve Jones, a geneticist at University College London, believes natural selection is not currently important to humans. He states that 500 years ago, a British baby had a 50 percent chance of living to adulthood. Now that baby's odds are 99 percent. However, Jones admits that natural selection is active in a few places. For example, the gene CCR5-32, which gives some resistance to HIV, is increasing in frequency in parts of Africa with a high AIDS rate.[10]

As changes in gene frequency cannot be detected except over generations, it is difficult to provide evidence of present-day natural selection in humans. However, we live in a rapidly changing physical and social environment, and we still comply with Charles Darwin's three criteria for natural selection.

The survival of premature babies could lead to an increase in an inherited tendency to give birth prematurely.[11] Caesarean sections could lead to less selection against larger babies.[12] The populations with the fastest growth rates are in South America, Africa, and Asia. The unique gene pools in these areas are growing faster than the gene pools in Europe or

North America. Many college students put off having children until later in life. Over generations, this could lead to a lower percentage of "college material" people. Women with Tourette's syndrome and attention-deficit/ hyperactivity disorder are less likely to attend college, thus increasing the probability they will produce children and grandchildren in a shorter period of time.[13]

It's possible that current unprecedented survival rates will result in their own kind of evolution. In *The Descent of Man,* Darwin discussed possible deleterious effects of welfare, charity, and medicine:

> With savages, the weak in body or mind are soon eliminated; and those that survive commonly exhibit a vigorous state of health. We civilized men, on the other hand, do our utmost to check the process of elimination; we build asylums for the imbecile, the maimed, and the sick; we institute poor-laws; and our medical men exert their utmost skill to save the life of every one to the last moment. There is reason to believe that vaccination has preserved thousands, who from a weak constitution would formerly have succumbed to small-pox. Thus the weak members of civilized societies propagate their kind.[14]

Gregory Cochran, adjunct professor of anthropology at the University of Utah in Salt Lake City, believes that "relaxed selection combined with a high mutation rate is probably causing gradual deterioration of many functions, especially disease defenses."[15] Another function that is likely to have deteriorated is predator avoidance. Selective pressure on humans by predators is almost unheard of.

Christopher Wills cites the case of severe combined immunodeficiency.[16] Researchers at the National Institutes of Health took white blood cells from two girls with the disease, gave the cells corrected copies of the defective gene, and transfused them back into the girls. The girls developed the ability to fight diseases that had formerly threatened them. But because the corrected genes are in their white blood cells only, and not in their eggs, their defective genes will be passed on. It is unlikely their future mates will have a copy of the same defective gene, but the girls' survival will increase the probability that sometime, generations from now, that gene will find another like itself and another diseased baby will be born who otherwise wouldn't have. Wills concludes, "Overall, though, the defective genes will have virtually no impact on the gene pool of our species, for they will simply join those billions of other mutant genes, new and old, that we are all collectively passing on to our children."

Wills alludes to an important point. The gene coding for severe combined immunodeficiency is not being selected *for* as a result of the gene therapy; rather, it is no longer being selected *against*. True reproductive discrepancy—people dying or not able to have children, or people able to have more children because of a particular gene—is hard to identify as a present-day force. Perhaps in one or two generations we will be better able to answer questions about what selective pressures were important in the year 2013.

TOMORROW

So where are we going? If the past is an indication of the future, we will continue to adapt genetically to changes in our environment, whether that environment is natural or created by us. We will continue to evolve as long as we reproduce sexually. Some futurists foresee us downloading our memories and personalities—our "souls" in the words of one writer—into machines that are superior in every way to our organic bodies.[17] If this happens, it will be debatable whether or not we are still undergoing organic evolution.

In considering future human evolution, there are a few forces that are interesting to examine: assortative mating, genetic engineering, germ-line genetic engineering, human augmentation, and changes in sociopolitical structure.

People have always been able to find mates with similar qualities because like people tend to live and work in similar places; however, the Internet has taken this a step further. The impact of Internet dating and other Internet-based means of meeting people is currently of limited influence, but their popularity and acceptance are growing. Dating could become dominated by nontraditional matchmakers that match people with similar traits more efficiently than meeting by chance. Being able to more precisely define required mate criteria could result in greater discrepancies in attractiveness, intelligence, and mental and physical health as people choose mates similar in these qualities.[18] However, sophisticated computer matchmakers might introduce people of dissimilar personality for the sake of compatibility.

Genetic engineering therapy, such as in the case of severe combined immunodeficiency discussed above, will continue to improve the lives of people with genetic disorders. While these changes are not heritable, there will no longer be a reproductive disadvantage for those who can afford the therapy and the defective genes will be preserved in the population.

If germ-line genetic engineering (that is, therapy that results in changes in genes that can be passed on) becomes common, natural selection will fall into the hands of mom and dad. I foresee parents and counselors discussing how much intelligence, beauty, aggressiveness, physical strength, outgoingness, leadership ability, and compassion they would like to see in their child-to-be. They will choose the sex and skin, hair, and eye color. The counselor will advise them on the compatibility of the genes they have chosen and perhaps will advise them on how compatible the child will be with the parents' friends' children. Parents might form unions so that communities of like children will grow up together and support each other optimally. These communities (not necessarily defined by geography) might become units of competition or cooperation. "Designed communities" might compete with each other in sports, business, cyber war, or street war. One group might emphasize wisdom and compassion, while another emphasizes analytical reasoning and strategy. After hundreds of generations, these communities might diverge genetically until they can no longer reproduce with each other without medical assistance. Designed babies will pass on their perfect genomes, and Mr. Darwin will have to call this evolution, possibly even speciation.

Some of this could make our present definition of evolution academic. For example, the addition of artificial chromosomes with genes for desirable traits would be heritable, but these traits would not be selected for because every generation could change the artificial chromosome as desired.[19] One can only imagine what Mr. Darwin would think.

Humans are already mechanically augmented. People today carry retinal electrodes, artificial limb actuators, cochlear implants, pacemakers, and artificial eye lenses and joints. Futurists such as J. Storrs Hall and Ray Kurzweil predict we will add to this list respirocytes (cell-sized robots that would carry far more oxygen than a red blood cell), microbivores (blood-born robots that kill anything pathogenetic), artificial winch cells in muscle, and artificial brain implants with almost unlimited long-term memory and additional reasoning.[20] Augmentations are not heritable, but they could lead to greater survivorship of a particular class of people.

Socio-economics affect evolution. Genocide, racism, foreign aid, access to technology, war, freedom and lack of it, policies on reproduction, and social conflicts such as illegal drugs and corruption have and will affect who survives and who reproduces. The absence of a reasonable justice system could lead to selection for the strong and the cruel. The development of a fair, efficient, and intelligent form of government could favor compassion.

It is possible that the rich will have greater access to assortative mating, genetic engineering, germ-line genetic engineering, human augmentation, and other technologies such as advanced traditional medicine and nutrition. This would result in a reproductive advantage of the haves over the have-nots.

2070 Again

Karen Armstrong, author of *A History of God*, believes humans are genetically hardwired to seek purpose in communion with something larger and more powerful than themselves.[21]

If this is true, the people of the year 2070 were about to answer their evolutionary calling. Humanity was to become one interconnected consciousness, and it would be the final fulfillment of a deep and sacred yearning. The earth's 40 billion awaited alone, with families, and with friends, in a worldwide cyber-handhold.

Joe didn't make it, nor did any of the other 40 billion. Humanity and its aids needed a little more understanding about quantum black holes and the side effects of the gravity wave human brain interface. With a quiet pop, all but the 3,000 who refused to participate collapsed where they stood, brain stems scrambled. Pithed by the singularity.[22]

The 3,000 survivors were emotionally shocked for several seconds, and they lived through a moment of sadness before they remembered their antidepressant software. Then, for the first time in decades, humans faced a real challenge. Questions of where they were going and how they were going to get there stimulated humanity in a way no one alive had ever experienced. And considering no one was very upset about the four billion perished, some wondered who they were. Were they themselves, or were they their software augmentations?

Most of the remaining population was not inclined to seek the same kind of universal spiritual fulfillment as the 40 billion had. With this personality type came independence and a tendency to not see one's individual life as quite so important. They still reproduced sexually, and their philosophical orientations would be passed on. Despite their best efforts, humans and their mechanical companions had failed to take evolution into their own hands. ❧

NOTES

1 Ward, P. 2009. The Future of Man—How will Evolution Change Humans? *Scientific American*, January.

2 McAuliffe, K. 2009. They Don't Make Homo Sapiens Like They Used To. *Discover*, March.

3 Ibid.

4 Hawks, J., E. Wang, G. Cochran, H. Harpending and R. Moyzis. 2007. "Recent Acceleration of Human Adaptive Evolution." *Proceedings of the National Academy of Sciences* 104 (52): 20753–8.

5 Begley, S. 2007. The New Science of Human Evolution. *Newsweek*, March.

6 Ruiz-Pesini, E., D. Mishmar, M. Brandon, V. Procaccio and D. Wallace. 2004. "Effects of Purifying and Adaptive Selection on Regional Variation in Human mtDNA." *Science* 303: 223–5.

7 Strohl, K. 2008. Lessons in Hypoxic Adaptation from High-Altitude Populations. *Sleep and Breathing*, May.

8 Ibid.

9 Stock, G. 2002. R*edesigning Humans: Our Inevitable Genetic Future.* New York: Mariner.

10 Douglas, K. 2006. Are We Still Evolving? *NewScientist*, March 11.

11 Imdtonson, R. 2008. Evolution Pressed. *NewScientist*, January 26.

12 Douglas, 2006.

13 McAuliffe, 2009.

14 Darwin, C. 1871. *The Descent of Man, and Selection in Relation to Sex*. New York: D. Appleton and Co.

15 Douglas, 2006.

16 Wills, C. 1992. Has Human Evolution Ended? *Discover*, August.

17 Katz, B. 2008. *Neuroengineering the Future. Virtual Minds and the Creation of Immortality.* Hingham: Infinity Science Press LLC; Hall, J. 2005. *Nanofuture*. Amherst: Prometheus Books; Kurzweil, R. 2005. *The Singularity is Near. When Humans Transcend Biology.* New York: Viking.

18 Douglas, 2006.

19 Garreau, J. 2004. *Radical Evolution.* New York: Doubleday; Bosveld, J. 2009. "Evolution by Sensing." *Discover* 30 (2): 58. Karen Armstrong was interviewed by Joel Garreau.

20 Hall, 2005; Kurzweil, 2005.

21 Garreau, 2004.

22 Kurzweil, 2005. The singularity, as described by Kurzweil, is a period of time in human evolution when technology accelerates at an advanced exponential rate. Kurzweil lists several principles which characterize the singularity. These are summarized by the following: 1) double exponential growth of information technology, 2) computer intelligence indistinguishable from human intelligence, 3) intelligent machines that can instantly share information and memories with all of civilization and can manipulate their own designs, 4) machines that are emotionally intelligent, 5) manipulation of matter at the molecular level, and 6) ultimately, the saturation of the universe with human intelligence. Kurzweil does not predict a date for the singularity (although well within 100 years is implied) and does not suggest it will happen all at once.

7

Darwin and Social Darwinism

The Use and Misuse of Science

Guy V. Beckwith

Sigmund Freud, the father of psychoanalysis, wrote that in the history of science there had been three great shocks to human pride.[1] The first was the Copernican revolution, which destroyed the idea that the Earth was at the center of the cosmos—thereby displacing the Earth's human inhabitants from the center as well. Copernicus's discoveries deflated human pretensions. The second historic shock we will turn to in a moment. The third—according to Freud—was Freud's own work, which demonstrated that humans were not the completely rational, self-determining creatures they thought they were; instead, they were prey to all manner of unconscious and irrational forces. In other words, humans were not masters of their own souls, let alone of the cosmos. I hope we can forgive Freud's ranking himself as the equal of Copernicus. Who knows: though he is out of favor at present, we may eventually conclude that his pioneering work did occur on that high level.

The second shock to human pride was, Freud claimed, the Darwinian revolution, because Darwin demonstrated that humans were not the result of a special creative act by a divine being, but were descended from the lower animals. Humans were not the divinely appointed lords of creation; they were biological organisms subject to the laws of evolutionary biology. This, Freud declared, was a shattering blow to our pride—such a blow that many people rejected Darwinism without a real hearing in the decades following the 1859 publication of *On the Origin of Species*. In fact, it was such a blow that some still reject the theory today, not because they dismiss all science or because they have carefully studied the theory as science, but merely because it hits them where they live. It strikes at a certain kind of human arrogance. There may be good reasons to reject or to revise Darwin's ideas, but those reasons are all too often lost in the rhetoric of wounded egotism.

Historically speaking, Freud was right: Darwinism was a shock to the collective psyche of Western civilization. James Burke puts it bluntly: "*The Origin of Species* hit the world like a bombshell . . ."[2] One of Darwin's fellow scientists, botanist Hewett Watson, wrote to him shortly after the publication of the book and clearly understood the revolutionary impact Darwin's theory would have for science and scientists:

> Your leading idea will assuredly become recognized as an established truth in science, i.e., "Natural Selection." It has the characteristics of all great natural truths, clarifying what was obscure, simplifying what was intricate, adding greatly to previous knowledge. You are the greatest revolutionist in natural history of this century, if not of all centuries.[3]

But the shock waves would not be limited to the world of "natural truths." As science historian Anthony Alioto tells us,

> Darwin's mechanism of evolution fostered one of the greatest revolutions in Western thought, for he dispensed with the ancient and honored doctrine of teleology, replacing it with continuous variation. His was a philosophical revolution as much as a biological one.[4]

Roland Stromberg, a noted historian of ideas, puts it even more strongly:

> Few areas were untouched by the Darwinian revolution . . . Religion, philosophy, the social sciences, literature and the arts—none would be the same. The impact of evolution caused a basic shift in the structure of all thought.[5]

It wasn't the idea of evolution in and of itself that was so revolutionary. Evolutionary theories were already on the scene and were championed by eminent thinkers such as Jean Lamarck in France, Johann Herder in Germany, and Darwin's own grandfather, Erasmus Darwin, in Britain. But these theories were speculative. They were essentially works of nature philosophy rather than of natural science. Darwin turned evolution into a scientific theory by basing it on careful observations, finely crafted hypotheses, and sustained (though often indirect) testing of those hypotheses. He gave the theory real force by working toward it inductively and by proposing a mechanism that both explained much of the raw data of biology and gave its fundamental principles a dynamism and inner coherence reminiscent of Newton's grand synthesis of physics and astronomy.

That mechanism—the heart of Darwin's theory and of its ultimate scientific success—was natural selection. Darwin argued that variations, or mutations, arise by chance in every generation.[6] Some may be beneficial to the organism in its particular environment as it competes with other organisms for the necessities of life. These positive changes tend to be preserved and passed on to later generations. Changes that are not beneficial, that do not improve an organism's ability to compete in the general struggle for survival, tend to drop out over time because they cripple or hamper the organism. Gradually, from generation to generation, descendants diverge from their parent stock. They become new varieties, new sub-species, new species, and ultimately whole new families and orders of creatures. The process is a blind, mechanistic one: chance provides the variations, and a chance congruence between mutation and the local environment provides the selection and the direction of development. With these mechanisms Darwin could explain design in nature strictly by natural laws and natural contingencies. He could explain design without a Designer, whether that designer be conceived of as God or as a spiritual principle in matter. Thus, Darwin could be the Newton of biology—demonstrating the laws and mechanisms of life—and at the same time he could be the David Hume of biology, magisterially refuting the traditional "argument from design." This was the revolution, and it threw Western thought into confusion because it was a frontal attack on so many cherished ideas and values.

The revolutionary force of his ideas made Darwin, on top of everything else, the Copernicus of biology. He took away the old stability, the old sense of a fixed and central place for humanity in the cosmos, and replaced it with a dizzying whirl of movement and change. For even preserved adaptations were not perfect and could be improved by new modifications and selections, and even a well-adapted organism or population could become unfit if the environment itself changed. Furthermore, species could and did become extinct. This terrible kind of death—death not just for one individual or another but for a whole type—gave people a cold and frightening sense of the power of chance and of change: "Blind chance, purposeless mechanism, tremendous biological waste—this was surely not the universe imagined by Natural Theology all the way back to Newton."[7] It was, instead, the darker world captured in Tennyson's chilling phrase, "Nature red in tooth and claw."

Since Darwin's theory excluded design from "outside," it rejected special creation. It was a terrible blow to human pride: there was no higher or lower realm, not in the old, strict, and reassuring sense. All adaptations were disconcertingly equal, and the human brain was in the last analysis no better an adaptation than the shark's teeth. Thus, humanity lost its exalted position in the cosmos.[8] The so-called "descent from the ape," destined to disturb and scandalize so many, was a side issue compared to the real implications of natural selection. Yet both themes had the same immediate effect: they sparked a furor of rejection and debate. I can only provide a few examples out of thousands. Adam Sedgwick, Cambridge professor of geology, insisted that Darwinism "would sink the human race into a lower grade of degradation" than it had ever known. (Given the depths to which humans had descended through the ages, this was saying a lot.) George Bernard Shaw, himself an evolutionist, nonetheless rejected Darwin's theory of natural selection. He wrote, "If it could be proved that the whole universe had been produced by such a selection, only fools and rascals could bear to live."[9]

The religious reaction was, predictably, the strongest. Many spokesmen for the churches, including the Church of England, saw Darwin's theory as an attack on scripture and on the church, an attack on the pivotal conceptions of spirit, soul, and human dignity, indeed, a radical reduction of the human to the animal. They saw Darwinism as an attempt to replace religious authority with an alternate authority, namely, a secularized science. In Darwin, they saw science giving aid and comfort to the enemy, i.e., to atheism and agnosticism, and in so doing threatening the very foundations of social order and civilization.[10]

Perhaps Bishop Samuel Wilberforce of Oxford stands as the best example of this initial, outraged reaction. He vehemently criticized the *Origin of Species* in an important intellectual journal and at the Oxford meeting of the British Association for the Advancement of Science in 1860. There, the bishop engaged in a spirited debate of the issues, not with Darwin himself, who avoided such encounters, but with the scientist's friend and champion, Thomas Henry Huxley—the man who became known as "Darwin's Bulldog." We are told that during this exchange the bishop contemptuously questioned Huxley as to whether he was descended from an ape on his grandfather's or his grandmother's side. This kind of insult was an accepted part of English debate at the time. Huxley turned to a companion and whispered, "God has delivered him into my hands." He then responded to the bishop by declaring that he would rather be descended from an honest ape than from a bishop who had brains but refused to use them. The audience exploded in laughter, and word spread that Huxley had crushed Wilberforce in open debate. This celebrated confrontation helped spur the acceptance of Darwinism among the scientific community and the scientifically educated not only in Britain but throughout the industrialized world.[11] The surprisingly rapid acceptance of the theory by scientists made its enemies initially all the more angry and vocal. Then, even some of them began to seriously consider Darwin's central idea of "descent with modification."

Why was Darwin's theory of evolution by natural selection accepted so rapidly by so many? If it was such a shock to traditional beliefs, why wasn't it ignored or marginalized? The Victorians were certainly not without resources when it came to suppressing or containing new and disturbing emotions and ideas. Why then did Darwinism, despite strong and very vocal opposition, flourish?

First of all, the theory spread quickly because it was, in the phrase used by many workaday scientists, "good science." *On the Origin of Species* sold out the first day of its release. It became a tremendous bestseller—for a book of hard science—as did *The Descent of Man*, which followed in 1871. Scientists and non-scientists alike devoured these books for the science. Many laypersons became amateur naturalists as a result.[12] Keep in mind that Darwin already had a solid reputation among professional naturalists as a result of his epic voyage on HMS *Beagle*. He returned from this adventure with an extraordinary collection of specimens, with a fascinating journal of his observations and preliminary conclusions, and with a new theory of the formation of coral islands. These won him fame among naturalists and election to important scientific societies decades before he published his masterwork. Though the rhetoric of "great" individuals and their deeds has been much abused and largely banished from academic writing, for some figures nothing else will do. Darwin was a great scientist: a patient observer, a meticulous recorder of observations and measurements, a rigorous thinker; and he had the imagination to synthesize his observations in a daring theory. Many scientists recognized these qualities, adopted the theory, and, drawing on their scientific authority and status, helped spread it among non-scientists.

The testimony of contemporaries helps illustrate Darwin's extraordinary powers of observation. Historian Timothy Ferris quotes Dr. Edward Lane, a friend of Darwin's who frequently accompanied him on his long walks. Lane tells us:

> No object in nature, whether Flower, or Bird, or Insect of any kind, could avoid his loving recognition. He knew about them all . . . could give you endless information . . . in a manner so full of point and pith and living interest, and so full of charm, that you could not but be supremely delighted, nor fail to feel . . . that you were enjoying a vast intellectual treat to be never forgotten.[13]

This kind of inspired empiricism, this love of precise observation, this painstaking care for facts, is undoubtedly part of what it means to do "good science."[14]

Darwin succeeded for other reasons as well. Not only did he amass a tremendous amount of evidence from many different fields—anatomy, animal husbandry, geology, paleontology—but all this evidence was superbly organized and carefully brought to bear on the hypothesis of evolution by natural selection. He also had a flair for written argument. A few readers found him dull, but intellectual historians praise his command of classical rhetoric and his ability to avoid technical jargon and address a wide audience.[15]

In addition, Darwin's work led to good science on the part of his disciples and followers. In other words, his was a fruitful theory, sparking new discoveries in biology, geology, paleontology, comparative anatomy, embryology, biochemistry, genetics, selective breeding in plants and animals, etc. Darwin himself stressed the importance of this continuation and testing of his approach. In a letter to Huxley, he says:

> I have got fairly sick of hostile reviews. Nevertheless, they have been of use in showing me when to expatiate a little and to introduce a few new discussions. I entirely agree with you, that the difficulties in my notions are terrific, yet having seen what all the Reviews have said against me, I have far more confidence in the general truth of the doctrine than I formerly had. Another thing gives me confidence, namely that some who went half an inch with me now go farther, and some who were bitterly opposed are now less bitterly opposed . . . I can pretty plainly see that, if my view is ever to be generally adopted, it will be by young men growing up and replacing the old workers, and those young ones finding that they can group facts and search out new lines of investigation better on the notion of descent than on that of creation.[16]

This is the voice of the scientist: He is stung by hostility but does not respond in kind. He thinks in terms of the future and does not fear the testing of his conceptions that the future will bring. There is a mildness to his reaction that is the hallmark of the open, humanistic brand of science celebrated by such figures as Jacob Bronowski, Carl Sagan, and Stephen Jay Gould. And Darwin was right about the future. New generations of scientists up to and including our own time have found continued inspiration for further research in Darwin's theory. With all of our accumulated knowledge of genetics and of prehistoric and historic development, the basic Darwinian edifice survives. Modern biology, with all its sub-disciplines, and modern biomedicine, with all its complexities and wonders, stand on the shoulders of Darwin. Thus, Darwin is recognized today as a scientific genius, and his theory as a triumph of human effort and discovery.

BUT—AND THIS IS a crucial point—Darwinism also spread for less noble reasons. As George Bernard Shaw put it: "Darwin had the luck to please anybody with an axe to grind."[17] That is, it was easy for those with a political or social agenda to adopt Darwinism as a way of advancing that agenda. And there were a lot of agendas around: nationalism, militarism, Marxism, laissez-faire capitalism, racism, imperialism, anti-Semitism, etc. The effort to apply Darwinism to human society is called "Social Darwinism." It was launched not by Darwin, who became quite skeptical of it, but by Herbert Spencer, a British philosopher who extended evolutionary theory to human life. He argued that "survival of the fittest"—a phrase Spencer introduced—was more than the mechanism of organic evolution: it was the basic law of society. It was the way humankind progressed. It must, then, be allowed free rein to weed out the unfit. Thus, most charitable or reform efforts designed to aid the poor were, in Spencer's view, misguided. As he declared in his first book, *Social Statics*:

> It seems hard that an unskilfulness which with all his efforts he cannot overcome, should entail hunger upon the artizan. It seems hard that a labourer incapacitated by sickness from competing with his stronger fellows, should have to fear the resulting privations. It seems hard that widows and orphans should be left to struggle for life or death. Nevertheless, when regarded not separately, but in connection with the interests of universal humanity, these harsh fatalities are seen to be full of the

> highest beneficence—the same beneficence which brings to early graves the children of diseased parents, and singles out the low-spirited, the intemperate, and the debilitated as the victims of an epidemic . . . to prevent present misery would entail a greater misery on future generations.[18]

Thus Darwin could be used—misused, rather—to help justify Spencer's philosophical commitment to laissez-faire capitalism: dog-eat-dog competition in the economic sphere was evolution in action.[19]

This might have remained an obscure philosophical conception but for the fact that Spencer became immensely popular. He "was close to being the most popular serious thinker of all time, to judge by the sale of his books." Contemporaries compared him to Thomas Aquinas and Descartes.[20] So when he applied Darwinian ideas to economics and society, many took notice.

One was William Graham Sumner, an American sociologist who championed Social Darwinism in the United States. He closely followed Spencer, using Darwinism to justify American "rugged individualism" and laissez-faire economics. To quote Sumner:

> It may shock you to hear me say it, but when you get over the shock, it will do you good to think of it: A drunkard in the gutter is just where he ought to be. Nature is working away at him to get him out of the way, just as she sets up her processes of dissolution to remove whatever is a failure in its line.[21]

Perhaps Sumner was not really vindictive, or at least not actively so; at other points in his writings he suggests a somewhat more charitable line. But given the long history of human irrationality, given what Hannah Arendt terms the "banality of evil," it is only a short step from his suggestion that the drunkard be left in the gutter to actively shooting the man to get him "out of the way" all the faster. For a time, Sumner's more passive form of Social Darwinism held sway.

The movement had a major impact in the United States, where millionaire industrialists such as Andrew Carnegie and John D. Rockefeller became enthusiastic advocates. Here's what Carnegie said about his first encounter with Social Darwinism:

> That light came in as a flood and all was clear. Not only had I got rid of theology and the supernatural, but I found the truth of evolution. "All's well since all grows better" became my motto . . . Man was not created with an instinct for his own degradation, but from the lower he had risen to the higher forms. Nor is there any conceivable end to his march to perfection . . . We accept and welcome, therefore, as conditions to which we must accommodate ourselves, great inequality of environment; the concentration of business, industrial and commercial, in the hands of a few; the law of competition between these, as being not only beneficial, but essential to the future progress of the human race.

John D. Rockefeller echoed these sentiments, though with a touch of traditional piety:

> The growth of a large business is merely the survival of the fittest. The American Beauty Rose can be produced in the splendor and fragrance which bring cheer to the beholder only by sacrificing the early buds which grow up round it. This is not an evil tendency in business. It is merely the working out of a law of Nature and a law of God.[22]

It is difficult to avoid a skeptical view of these grandiose pronouncements. They obviously did not flow from dispassionate scientific analysis. Great industrialists and financiers were enthusiastic about Social Darwinism because, in their eyes, the doctrine proved that they were wealthy and powerful because they were the most fit, the most highly evolved, the most deserving, and that the poor—even those made poor by the deliberate actions of a Rockefeller or a Carnegie—were the unfit, who therefore deserved to be poor. This wasn't exactly a new position in America or in Britain, because the works of Adam Smith on laissez-faire capitalism had already been misread to serve this kind of conclusion. But it wrapped laissez-faire up in the borrowed robes of hard science and led to conclusions even more cruel than the old economic theory.

I am not trying to deny that Carnegie, Rockefeller, and many of the other leaders of American industry engaged in some praiseworthy works of philanthropy. It is to their credit that they weren't always consistent in their embrace of Social Darwinism. But they would have happily agreed with Sumner, who announced: "Millionaires are a product of natural selection . . . Let it be understood that we cannot go outside this alternative: liberty, equality, survival of the fittest; not liberty, equality, survival of the unfittest."[23]

All this was bad enough, and there was worse to come, in the United States and in many other countries. But the most frightening version of Social Darwinism appeared in Germany. The young scientist Ernst Haeckel (1834–1919), professor of zoology and comparative anatomy at the University of Jena, became the primary exponent of Darwinism in Germany. He gleefully accepted the Social Darwinist principle that humans were part of nature and subject to natural law, including the law of the survival of the fittest. But he went much farther. He argued for a direct, literal, and militant application of the laws of evolutionary biology to human life and society. He believed that Darwinian science proved German racial superiority. The German form of racism antedated Haeckel and Darwin, but Haeckel now gave it fake scientific credentials. In doing so, he became one of the forefathers of Nazi ideology. His version of Social Darwinism influenced all levels of German society. He taught that some races were more highly evolved than others and thus worth more, and that some individuals were more highly evolved and so their lives were "naturally" more valuable. He rejected free will as unscientific; thus, he rejected liberalism and democracy.

It is ironic: British and American Social Darwinism accepted individualism and democracy as consistent with Darwin, while German Social Darwinism rejected individualism for immersion in society, and rejected democracy for obedience and command—again, all in the name of Darwin. Haeckel and his followers declared that natural selection merely pointed us in the right direction: the primary business of the state and of society was eugenics or artificial selection—politics as applied biology. They advocated infanticide for sickly or deformed children and euthanasia for the insane, for habitual criminals, and for people with incurable diseases. They believed that we should breed human beings to improve the race. A surprising number of people, including many scientists, saw these startling proposals as reasonable and, indeed, as the fruit of the latest and best scientific thinking.[24]

Adolph Hitler took up the torch from Haeckel and his disciples. Hitler became a fanatical Social Darwinist and put the most cold-blooded of the ideas of the eugenics movement into effect in the Nazi state. During the short period of Nazi rule, women of supposedly "pure" German stock were urged to mate with young, healthy German soldiers, outside of marriage, as a way of improving the German race (which was, according to Nazi theory, already a super-race, biologically and in every other way). Those judged "unfit" were routinely mocked, exploited, and eliminated.

Eugenics in the hands of the Nazis gave birth to a nightmare. The infamous death camps of World War II were in part a way of keeping those groups and races the Nazis detested from propagating themselves. Men like Goebbels and Himmler were enthusiastic about the idea that people they deemed genetically inferior would not be allowed to reproduce. Sterilization experiments were carried out on a large scale at some of the camps. Adolf R. Pokorny, a notorious Nazi SS doctor, wrote Himmler that his research group had discovered a plant derivative that induced lasting sterility in the prisoners: "The thought alone that the three million Bolsheviks now in German captivity could be sterilized, so that they would be available for work but precluded from propagation, opens up the most far-reaching perspectives."[25] (That last phrase was Hitlerian bureaucratese for ecstatic approval.)

Let me reiterate: racism, rabid nationalism, and ethnocentrism were already well-entrenched in Western civilization. Darwinism didn't create them. But Darwinism was appropriated by them and used to give them a pseudo-scientific legitimacy. We can see the fallacy involved if we look at the language of Haeckel:

> Evolution and progress stand on the one side, marshaled under the bright banner of science; on the other side, marshaled under the black flag of hierarchy, stand spiritual servitude and falsehood, want of reason and barbarism, superstition and retrogression . . . Evolution is the heavy artillery in the struggle for truth; whole ranks of dualistic sophisms fall before [it] as before the chain shot of artillery.[26]

This is the language of propaganda, of ideology, of war—not the language of reason, of criticism, of genuine concern for the truth. *And this man was a trained scientist.* Hitler's rhetoric was even worse—yet, questioned, he would have insisted that National Socialist eugenics was firmly based in science, in Darwinian evolution. But Darwin himself was a man of charity and benevolence, who felt the need to apologize in his autobiography for not devoting all of his time to philanthropy.[27] Darwin and his extended family were also deeply committed to the movement to abolish slavery. While voyaging in the *Beagle*, Darwin expressed outrage at slave-owning Europeans and the cruelties they visited on their human victims. Yet Nazi Social Darwinists reintroduced slavery and institutionalized cruelty. This is perhaps the most painful irony of all.[28]

What, then, can we learn about science as a way of thinking, and about the uses of science in human life and history, from this profoundly disturbing devolution of ideas? I am reminded of Jacob Bronowski's film *Knowledge or Certainty*, one of the best works in his groundbreaking Ascent of Man series. In that film, Bronowski argues that the horrors of the 1930s and of World War II were not the tragedy of science but the tragedy of humankind. By this, I take it he means that the death camps and atomic bombs were not a tragedy of science as a method, as a system, and as an intention. In other words, it wasn't the tragedy of science at its best. But it was a tragedy of science in another sense, as a human activity, subject to the manifold errors, confusions, and cruelties of human life. It stands as a warning: even the best science can be diluted and misused. Often this dilution and diversion is the work of non-scientists, of politicians and financiers and ordinary people in the street. But sometimes it is the work of scientists themselves, in moments of ignorance, twisted enthusiasm, and moral failure. The clear historical fact that science can be misused even by scientists should remind us that what Shaw called "moral passion," the all-consuming desire to discover and actually do what is right, requires a vigilant, continuous interrogation of self and situation. Not only is the unexamined life not worth living: armed with the physical and theoretical powers of the sciences, it is an invitation to catastrophe.

We can learn something else as well. In the sciences—human as well as natural—context is everything. Principles and conclusions that are perfectly valid within their proper context become invalid when ripped out of it. When Einstein's theory of relativity, one of the high watermarks of human intelligence, becomes the catchphrase "all is relative," its real meaning disappears, and a terrible closing of the mind results. When Heisenberg's uncertainty principle becomes the smug assertion that "all is uncertain," knowledge becomes impossible, including the knowledge that gave rise to the principle in the first place. When evolution by natural selection, Darwin's historic contribution to biology, is forced from its context and artificially grafted to politics or economics, we see the same dismal result.

This is not to say that science cannot contribute to other areas of life and to our daily lives. But it must be real science, understood first within the limits and qualifications of its origin. We must use science with wisdom and a healthy humility, or we destroy its capacity to give us understanding and real progress.

Ironically, Social Darwinists managed to reintroduce the all-too-human arrogance that the shock of Darwinism had, in principle, exorcized. Yet the real lesson of Darwinism is greater humility, grounded in a sense of kinship with all life and an acceptance of the value, to life and to human life, of diversity. This lesson is all the more important today, for Social Darwinism is not dead. It is still very much alive, though it does not usually go under its old name. Any person who claims that everyone who is poor deserves to be poor, that other races or ethnic groups are inherently inferior, that power and wealth are sure signs of fitness, of goodness, of value to society, is very much a part of this old and discredited movement.

Such phrases misuse scientific terms and discoveries and make a mockery of the lives and labors of genuine scientists.

Certainly we must have science, and use science. Otherwise we will cripple ourselves in the art of living. But we must learn to use science with, as Bruno Bettelheim suggested, an "informed heart."[29] This is admittedly no easy task. We have had much more historical (and personal) experience inflating false pride and pandering to it. But if we pay careful attention to that history, painful as it has been, and to all the ways in which our egotism and irrationality have perverted the most promising of our intellectual and technical achievements—if, in other words, we foster a heightened historical and philosophical awareness of our use and misuse of science—then we stand a much better chance of transforming science, and all forms of knowledge, into tools in service to life. And that is what Charles Darwin really wanted.[30] ❧

Notes

1 Freud, S. 1916. *Introductory Lectures on Psychoanalysis*. trans. and ed. J. Strachey. New York: W. W. Norton and Company, 1977, 284–5.

2 Burke, J. 1986. *The Day the Universe Changed*. Boston: Little, Brown and Company, 260.

3 Eiseley, L. 1956. "Charles Darwin," *Scientific American*, February. Quoted in Eiseley (1956).

4 Alioto, A. 1987. *A History of Western Science*. Englewood Cliffs: Prentice-Hall, 276.

5 Stromberg, R. 1994. *European Intellectual History Since 1789*, 6th ed. Englewood Cliffs: Prentice Hall, 109.

6 Darwin, C. 1859. *On the Origin of Species by Means of Natural Selection*. Mineola: Dover Publications, 2006, 83. Darwin sees mutations as arising from complex and (in his time) largely unknown causes—thus they aren't a matter of pure chance. But they are emphatically not the result of some preordained design. Throughout his published and unpublished writings he stresses the role of chance and of accident in biological evolution, and later researchers tied this theme to mutation even more explicitly.

7 Alioto, 1987.

8 Abrams, M. 1974. The Victorian Age. In *The Norton Anthology of English Literature*, vol. 2, 3rd ed. New York: W. W. Norton and Company, 882.

9 Stromberg, 1994. Quoted in Stromberg (1994).

10 Houghton, W. 1957. *The Victorian Frame of Mind*. New Haven: Yale University Press, 58–9.

11 Ferris, T. 1988. *Coming of Age in the Milky Way*. New York: Morrow, 244–5; Mason, S. 1962. *A History of the Sciences*. New York: Collier Books, 422; Eiseley, 1956; Gopnik, A. 2009. *Angels and Ages: A Short Book About Darwin, Lincoln, and Modern Life*. New York: Alfred A. Knopf, 176.

12 Burke, 1986.

13 Ferris, 1988; Eiseley, 1956. Quoted in Ferris (1988).

14 Whitehead, A. 1925. *Science in the Modern World*. New York: New American Library, 1954, 2–3. Darwin's powers of observation are brilliantly evoked in Gopnik (2009), 64–8 and 72–4.

15 Darwin, 1859; Stromberg, 1994; Gopnik, 2009. See Michael T. Ghiselin's introduction to Darwin (1859).

16 Henahan, J., ed. 1975. *The Ascent of Man: Sources and Interpretations*. Boston: Little, Brown and Company, 207. Quoted in Henahan (1975)

17 Burke, 1986. Quoted in Burke (1986).

18 Spencer, H. 1851. *Social Statics, or The Conditions Essential to Human Happiness Specified. Reprints of Economic Classics*. New York: Augustus M. Kelley, 1969, 323.

19 Mason, 1962.

20 Stromberg, 1994.

21 Sumner, W. 1963. The Forgotten Man. In *Social Darwinism: Selected Essays of William Graham Sumner*. Englewood Cliffs: Prentice-Hall, 122.

22 Burke, 1986. Quoted in Burke (2006).

23 Ibid.

24 Stein, G. 1988. "Biological Science and the Roots of Nazism." *American Scientist* 76 (1): 53–57.

25 Shirer, W. 1960. *The Rise and Fall of the Third Reich*. Greenwich: Fawcett Publications, 1275. Quoted in Shirer (1960).

26 Gould, S. 1977. *Ever Since Darwin*. New York: W. W. Norton and Company, 217. Quoted in Gould (1977).

27 Darwin, C. 1858. *The Autobiography of Charles Darwin*. ed N. Barlow New York: W. W. Norton and Company, 1958, 95.

28 Desmond, A. and J. Moore. 2009. *Darwin's Sacred Cause: How a Hatred of Slavery Shaped Darwin's Views on Human Evolution*. New York: Houghton Mifflin Harcourt; Gruber, H. and P. Barrett. 1974. *Darwin on Man*. New York: E. P. Dutton and Company; Gould, S. 1992. "The Most Unkindest Cut of All." *Natural History*, June: 8–11. For Darwin's lifelong and passionate commitment to the anti-slavery movement and its belief in universal human brotherhood, see Desmond and Moore (2009). Darwin's ourage at European mistreatment of African slaves is documented in Desmond and Moore's *Darwin* (1992). His enlightened view of race is explored in Gruber and Barrett (1974) and in Gopnik (2009). Darwin was, understandably, not completely immune to the biases and eugenic anxieties of his time (Desmond and Moore, 2009), but he was no racist. The rank injustice of connecting Darwin to Hitlerian Social Darwinism is clearly demonstrated in Gould (1992).

29 Bettelheim, B. 1960. *The Informed Heart: Autonomy in a Mass Age*. Glencoe: The Free Press.

30 Marcel, G. 1971. *Man Against Mass Society*. Chicago: Henry Regnery Company, 55-6; Jonas, H. 1984. *The Imperative of Responsibility: In Search of an Ethics for the Technological Age*. Chicago: University of Chicago Press, ix–xi.

8

Darwin's Legacy in Psychology

Lewis Barker

Psychology is defined as the study of mind and behavior. Charles Darwin (as well as other 19th-century naturalists) was interested in and wrote compellingly about both. He speculated on similarities and differences of human and animal minds and on the nature of the human mind and human consciousness. Given that a defining feature of the human mind is to contemplate itself, he wondered about the ways human minds differ from the minds of other species. He wrote about human and animal instincts, human development, and human and animal emotions. Darwin went so far as to advocate the research direction of a future psychological science:

> In the distant future I see open fields for more important researches. Psychology will be based on a new foundation, that of the necessary acquirement of each mental power and capacity by gradation. Much light will be thrown on the origin of man and history.[1]

Darwin's publications changed almost everything of interest to psychologists of his day and continue to influence contemporary psychology and neuroscience. In a letter to Darwin, after reading *Origin of Species*, Francis Galton captured this sea change in thinking:

> I have laid it down in the full enjoyment of a feeling that one rarely experiences after boyish days, of having been initiated into an entirely new province of knowledge which, nevertheless, connects itself with other things in a thousand ways.[2]

It is this connecting "with many other things" that captures Darwin's influence on many emerging sciences in the 19th and early 20th centuries.

Pre-Darwinian Philosophy and Psychology

With roots in both philosophy and biology, psychology as a formal discipline is a latecomer to the sciences, existing only from the late 19th century. Darwin's writings coincided with the birth of formal psychological inquiry. Earlier philosophers had set the occasion for a nature or nurture pseudo-argument that concerned many 20th-century psychologists.[3] In Western thought, John Locke (1632–1704) emphasized the role of culture in making humans human. Locke's *An Essay Concerning Human Understanding* discussed the origin of ideas. He posited that ideas resulted from experiences in a lifetime. A human came into the world as a *tabula rasa*—a "blank slate" upon which individual experiences were writ. For Locke, the formation of complex ideas was accomplished through the association of simpler ideas built on raw sensory experiences.[4] Other than reference to said raw sensory experiences, implying the organs of eyes and ears, Locke downplayed the biological nature of humans.

In response to Locke, Gottfried von Leibniz (1646–1716) proposed that humans had an innate intellect. Leibniz was not anticipating contemporary ideas of innate brain modules; indeed, neither Locke nor Leibniz addressed brain functioning in their speculations about the origins of human intellect. Likewise, Immanuel Kant (1724–1804) was assuredly not thinking of brain mechanisms when he asserted that humans are ". . . endowed with a priori knowledge." None of these philosophers were interested in or considered the natural history of humans or the organ of emotion and thought.

Prior to Darwin's writings, the central role of brain functioning in contemporary psychology is better traced to Rene Descartes's (1596–1650) dualism, and to Franz Gall's (1758–1828) phrenology. While acknowledging human biology, both ideas were limited by lack of methodology and meaningful empirical observation. Descartes, of course, proposed qualitative differences between humans and other animals. During Darwin's productive years (and somewhat prior to psychology becoming a distinct discipline), neurophysiology appeared: Charles Bell (1774–1843) and Francois Magendie (1783–1885) distinguished between sensory and motor nerve functioning, and Johannes Muller (1801–1855) proposed the doctrine of specific nerve energies. Yet to be discovered were the communicative centers of the human brain by Paul Broca (1824–1880) and Carl Wernicke (1848–1905).

Darwin's Contemporaries and Founders of Psychology

Although not considered a psychologist, Francis Galton (1822 –1911) made important and lasting contributions to the field. As noted above, Galton was heavily influenced by *Origin of Species* and 10 years later wrote *Hereditary Genius* (1869).[5] Indeed, though he is considered an anthropologist, eugenicist, geographer, and meteorologist, Galton, not Wilhelm Wundt or William James (see below), was also the first modern psychologist. Recognizing the importance of (what are now called) genes and environment as co-determinants of human behavior, he coined the phrase "nature versus nurture." He was a psychometrician—a statistician who defined probability, correlation and regression and who applied the bell curve to human behavior.[6]

Galton's somewhat invisible role in the history of psychology is even less understandable in light of his recognition of the importance of individual differences. Fascinated with Darwin's discussion of artificial selection, Galton's insight was that the biological variability necessary for a theory of natural selection could be applied to a better understanding of the human condition. Individuals vary both biologically and psychologically. Galton advocated the study of identical and fraternal twins to help separate the interactive effects of nature and nurture.[7] Given that the study of individual differences defines modern psychology, Galton's impact is arguably more important than that of Wundt, the "first" psychologist.

Darwin as Psychologist

The "history of psychology" also includes Darwin's behavioral interests and his influence on the training of early psychologists. Psychological thinking is apparent in three of Darwin's seminal writings: *Origin of Spe-*

cies in 1859 (humans and other animals are biologically continuous); *On the Expression of Emotions in Man and Other Animals* in 1872 (humans and other animals are behaviorally and psychologically related)[8]; and *The Descent of Man, and Selection in Relation to Sex* in 1871 (mate choice via attraction complements natural selection in producing each generation).[9]

Although most psychologists consider Darwin's later written works to be his main contributions to psychology, I am not alone in arguing that *Origin of the Species* was more influential.[10] Darwin provided a new creation myth, based in nature and supported by empirical evidence, that integrates findings from many sciences. If humans are discontinuous with other animals, a separate, *sui generis* psychology can be argued. Human cognitive and affective attributes could take any form, unconstrained by a brain designed over millions of years through processes of natural and sexual selection. This is not to downplay the role of environment (read culture) in shaping each human's brain; rather, the point is that our evolved brain is the starting point for a human psychology. Donald A. Dewsbury has suggested, for example, that Darwin's writing about instinctive behavior moved it ". . . from the supernatural and theological to the natural."[11]

Darwin wrote about instinct? Imagine taking the following quiz: Which of the following are the most unlikely chapter titles in *Origin of Species*? "Variation Under Domestication," "Variation Under Nature," "Struggle for Existence," "Natural Selection," "Laws of Variation," "Difficulties on Theory," or "Instinct"? These are titles of the first seven chapters of Darwin's 14-chapter book. To be fair, the struggle for existence may imply behavior, but there is no mistaking the behavioral content (and discussion of "mental actions") of chapter seven, entitled "Instinct." Here is what Darwin says to introduce his conception of instinct:

> I must premise, that I have nothing to do with the origin of the primary mental powers, any more than I have with that of life itself. We are concerned only with the diversities of instinct and of the other mental qualities of animals within the same class.I will not attempt any definition of instinct. It would be easy to show that several distinct mental actions are commonly embraced by this term; but every one

> understands what is meant, when it is said that instinct impels the cuckoo to migrate and to lay her eggs in other birds' nests. An action, which we ourselves should require experience to enable us to perform, when performed by an animal, more especially by a very young one, without any experience, and when performed by many individuals in the same way, without their knowing for what purpose it is performed, is usually said to be instinctive.[12]

Darwin could not have been aware of the future difficulties generations of ethologists, comparative psychologists, and behavioral biologists faced in characterizing "instinctive behavior." For example, to what extent is behavior dictated by "selfish genes"?[13] What is the role of environment in "instinctive behavior"?[14]

Comparative Psychology. Though Darwin wrote about behavior in *Origin of Species*, the publication of *Expression of Emotions* founded comparative psychology—the study of the motivation and emotion in guiding the behavior of animals. But *Origin of the Species* provided the context. Human motivation and emotions exist in response to the selective pressures that shaped them. The endless varieties of human motivation and emotion can be viewed as "special processes that enhance fitness in certain situations . . ."

> Fear motivates escape from danger; anger motivates attack; joy motivates continuing on the present course, or, if the object has been obtained, ceasing to strive for it; disgust motivates avoidance, vomiting, and more metaphorical expulsion; interest motivates exploration; lust motivates seduction and sexual intercourse; sorrow motivates calling for help or giving up on fruitless endeavors, and so on.[15]

Sexual Selection. Darwin's theory of sexual selection influenced social psychology only late in the 20th century. Writing in *Descent of Man*, Darwin wondered why so many species of female fish and birds were typically drab compared to their male counterparts and why male peacocks would squander so much energy to produce beautiful, but otherwise useless, tail feathers. His theory, sexual selection, was again deceptively simple—"the advantage which certain individuals have over others of the same sex and species, in exclusive relation to reproduction."[16] David Buss identifies the two main components of this theory: same-sex competition (males competing with males and females competing with females for mates), and mating preferences (how females and males choose with whom to mate).[17]

DARWIN AS NEUROSCIENTIST

It is difficult to resist the tendency on this occasion of "looking back" to overstate the influence of Darwin's ideas on the subsequent development of the biological and psychological sciences. For example, few would think of Darwin as a neuroscientist or as having influenced current thinking in the neurosciences. However, consider the following paragraph from *Descent of Man*:

> As the various mental faculties gradually developed themselves the brain would almost certainly become larger. No one, I presume, doubts that the large proportion which the size of man's brain bears to his body, compared to the same proportion in the gorilla or orang, is closely connected with his higher mental powers. We meet with closely analogous facts with insects, for in ants the cerebral ganglia are of extraordinary dimensions, and in all the *Hymenoptera* these ganglia are many times larger than in the less intelligent orders, such as beetles. On the other hand, no one supposes that the intellect of any two animals or of any two men can be accurately gauged by the cubic contents of their skulls. It is certain that there may be extraordinary mental activity with an extremely small absolute mass of nervous matter: thus the wonderfully diversified instincts, mental powers, and affections of ants are notorious, yet their cerebral ganglia are not so large as the quarter of a small pin's head. Under this point of view, the brain of an ant is one of the most marvelous atoms of matter in the world, perhaps more so than the brain of a man.[18]

Here Darwin anticipates Jerison's regression of brain mass-by-body mass in comparative species,[19] and as well foreshadows brain-behavior relationships which define the neurosciences.[20]

Darwin's Impact: 19th-Century American Psychologists

Beginning in 1879, a number of Americans trained as post-doctoral students at the laboratory of Wilhelm Wundt in Leipzeig, Germany. Wundt is credited as the first psychologist whose theories were informed by laboratory-generated evidence. A close contemporary, William James, also had a laboratory at Harvard at this time. It is noteworthy that the 1879 founding date follows by 20 years Darwin's seminal publication of *Origin of Species* and, by less than a decade, his *Expression of Emotions* and *Descent.* G. Stanley Hall, James McKeen Cattell, and Frank Angell were among the Americans who trained with Wundt. All returned and started psychological laboratories in American universities. Another, a Brit, Edward Titchener, migrated to America as a professor of psychology at Cornell.[21] Is it fair to ask to what extent Darwin's thinking influenced Wundt, James, and others, and whether this influence was transmitted to the contingent of professors who studied with them?

These questions are not easily answered regarding Wundt. His research program, which became known as structuralism, was concerned with the consciousness of perception. The method Wundt developed, called introspection, was a conscious attempt to mimic the successes of the emerging fields of physiology and chemistry. He aimed to produce a "chemistry" of the mind by examining responses to sensory stimuli.[22] In the United States, structuralism was championed for several decades by Titchener, but this school of psychology proved to be short-lived. Natural history and natural selection (and, for that matter, sexual selection and comparative animal behavior, which came with Darwin's later publications) influenced neither Wundt's nor Titchener's structuralism. In addition, Wundt and Titchener were not especially interested in individual differences; they assumed that "properly trained" human subjects gave similar responses in laboratory experiments. Structuralism had no lasting influence and did not thrive in American universities.

But in reconstructing such historical perspectives, one should always wonder about the zeitgeist. Was Darwinian thinking in the air? Was Wundt influenced by Darwin independently of the development of structuralism? Perhaps. Writing in 1892, well after all of Darwin's works were published, Wundt wrote in *Ethics: An Investigation of the Facts and Laws of the Moral Life*:

> Among the qualities thus developed by natural selection are the social instincts. Thus man is undoubtedly a social animal, distinguished from the lower animals only by his capacity for reflection. Even his simian ancestors apparently possessed the same instinct.[23]

It is likely Wundt's thoughts in 1892 did not reflect his thinking during the time he trained American psychologists in structuralism. They do reflect the evolutionary thinking of James at Harvard and Angell of the Chicago school, who developed ideas that were later called functionalism. What is clear is that at some point Darwin's thinking did influence Wilhelm Wundt, the "first" psychologist.

AMERICAN "FUNCTIONALISM"

In contrast with Wundt, there is no question of Darwin's influence on William James (1842–1910). The umbrella term "functionalism" was defined by James as the study of the usefulness of consciousness and the utility of behavior (rather than its contents). James's psychology, then, was based on adaptation during an organism's lifetime, analogous to Darwin's concept of a species's adaptations throughout geological time.

James wrote the first (and, some would argue, the best) widely adopted textbook of psychology, *Principles of Psychology*, in 1890. James incorporated Darwinian thinking from the first chapter ("Functions of the Brain") to the last chapter—1,400 pages later—in which he discusses "the origins of instincts." *Principles* took 12 years to publication. Writing about evolution, brain, and behavior, James is clearly influenced by Darwin:

> All nerve centres have then in the first instance one essential function, that of "intelligent" action. They feel, prefer one thing to another, and

> have "ends." Like all other organs, however, they evolve from ancestor to descendant, and their evolution takes two directions, the lower centres passing downwards into more unhesitating automatism, and the higher ones upwards into larger intellectuality. Thus it may happen that those functions which can safely grow uniform and fatal become least accompanied by mind, and that their organ, the spinal cord, becomes a more and more soulless machine; whilst on the contrary those functions which it benefits the animals to have adapted to delicate environing variations pass more and more to the hemispheres, whose anatomical structure and attendant consciousness grow more and more elaborate as zoological evolution proceeds.[24]

What is important to recognize in James's thinking 130 years ago is how prescient he was in anticipating the correspondences of brain, mind, and behavior within an evolutionary context. The "intelligent" action of nerves can be restated as the perfect adaptive fit of receptors and motor nerves—evolved reflexes. Brain stem and spinal cord are overlain by cerebral hemispheres which make possible civilized behavior in humans.

James's textbook addressed consciousness, memory, learning, perception, and development. He anticipated and defined the confluence of traditional concerns of psychology from an evolutionary perspective:

> The distribution of consciousness shows it to be exactly such as we might expect in an organ added for the sake of steering a nervous system grown too complex to regulate itself . . . Consciousness . . . has in all probability been evolved, like all other functions, for a use—it is to the highest degree improbable a priori that it should have no use.[25]

Darwin's seminal influence in the rapid development of comparative psychology can be seen in works published in the late nineteenth and early twentieth centuries: George Romanes's *Animal Intelligence* (1881), and *Mental Evolution in Animals* (1883); C. Lloyd Morgan's article on animal intelligence (1882) and *An Introduction to Comparative Psychology* (1894); and in Margaret Washburn's *The Animal Mind: A Textbook of Comparative*

Psychology (1908).[26] The question of human-animal mental continuity continues to the present. C. Lloyd Morgan in particular admonished animal researchers to be leery of anthropomorphism, which was a concern of all the foremost researchers. His oft-quoted "Morgan's Canon" states: "In no case is an animal activity to be interpreted in terms of higher psychological processes, if it can be fairly interpreted in terms of processes which stand lower in the scale of psychological evolution and development."[27]

"Morgan's Canon" is often interpreted as every scientist's interest in urging parsimony in the interpretation of observations. Later writings by Morgan suggest an alternative explanation. In *Emergent Evolution* (1921), Morgan proposes a concept of evolution that begins with matter, continues "upward" through all life forms, to become "consciousness"—which in humans may be enjoyed at a supra-reflective level. But to achieve this, he argues, humans must intuitively feel one with God in substance. Hence, only humans, and only some humans, can be fully conscious.[28]

Is this a theistic-inspired human-animal divide? Some of Morgan's contemporaries thought so. William McDougall was explicit in his objections: "If we turn to Lloyd Morgan's *Emergent Evolution*, we find no mention of Darwin or of Lamarck, of variation, mutation, or selection. He frankly asks: what makes emergents emerge? And his answer is: the directive activity of God."[29] Not only does Morgan's conjecture lack parsimony, a reexamination of his canon suggests a lack of symmetry not apparent on a first reading. Should we, as Darwinists, not also apply "animal characteristics" to humans if such processes stand lower in the scale of psychological evolution and development?

Morgan was not alone in his apparent attempt to salvage human dignity within Darwin's new universe of natural history. Such thinking also may have inspired Pierre Teilhard de Chardin's (1881–1955) idea of divinely guided evolution toward an omega point. It remains difficult to reconcile theistic beliefs with an evolution guided only by winners and losers, and an extinction record that favors the losers.

COMPARATIVE PSYCHOLOGY: LEARNING AND BEHAVIOR

An outgrowth of Romanes's and Morgan's interest in animal minds, investigations of how animals learn in laboratories became an important aspect of American functionalist psychology during the 20th century. Edward Thorndike (1874–1949), a student of William James, devised a "puzzle box" to measure a cat's ability to learn an escape response.[30] He conceptualized animal learning as instrumental to survival:

> The most important of all original abilities is the ability to learn. It, like other capacities, has evolved. The animal series shows a development from animals whose connection-system suffers little or no permanent modification to animals whose connections are in large measure created by use and disuse, satisfaction and discomfort.[31]

Thorndike proposed a "law of effect" based on innate biological tendencies that predicts how new behaviors are acquired:

> Of several responses made to the same situation, those which are accompanied or closely followed by satisfaction to the animal will, other things being equal, be more firmly connected with the situation, so that, when it recurs, they will be more likely to recur; those which are accompanied or closely followed by discomfort to the animal will, other things being equal, have their connections with that situation weakened, so that, when it recurs, they will be less likely to occur.[32]

Others (including America's most famous behaviorist, B. F. Skinner) have pointed out the close analogy of Thorndike's law of effect to Darwin's principle of natural selection.[33] Unreinforced responses "die" and reinforced responses are adaptively selected to "live" and be repeated. Even Pavlov acknowledged the adaptiveness of both unconditioned and conditioned responses:

> It seems obvious that the whole activity of the organism should conform to definite laws. If the animal were not in exact correspondence with its environment it would, sooner or later, cease to exist . . . The animal must respond to changes in the environment in such a manner that its

responsive activity is directed towards the preservation of its existence.[34]

Sexual Selection and Evolutionary Psychology

Darwin's sexual selection theory is the cornerstone of a subfield of psychology called, appropriately enough, evolutionary psychology. In *The Mating Mind,* evolutionary psychologist Geoffrey Miller uses sexual selection theory to describe courting tactics of humans.[35] He suggests that men parade muscles, music, attention-getting antics, humor, intelligence, and power (including financial wherewithal), and that women do the choosing.[36]

Language, Music, and Conceptual Issues

Perhaps the greatest challenge facing psychologists who attempt to understand the human condition from an evolutionary perspective is the gap from other animals in language, music, mathematics, and other conceptual abilities. The difficulty, of course, is the absence of evidence for an evolution of these psychological experiences in humans.[37] Such questions are of too broad to be addressed here. I will reference only two very different approaches. First, this volume contains a chapter by Jeff Katz et al. that describes a history of how comparative psychologists have studied cognitive processes in animals.[38] Here Katz et al. describe their research on conceptual learning by pigeons and monkeys.

The second example comes from Merlin Donald.[39] In his book *A Mind So Rare,* Donald proposes three stages in the evolution of human consciousness. The first (and most ingenious) stage is what he calls *mimesis* (from the Greek "to imitate, reproduce, or represent")—the human ability to gesture, act, practice, rehearse, imitate, and perfect. Nonhuman animals, he argues, do not have the mind or brain to engage in these behaviors. *Mimesis* is characterized by shared attention (between teacher and student, parent and child, etc.), necessary for acquiring skill in language, dance, music, art, tool-making, and so forth. (Others have characterized shared attention in terms of empathy, "mind reading," and a theory of mind.) *Mimesis* underlies the important role of imitation in acquiring new human abilities.

Donald's second stage of consciousness, language, is characterized as shared consciousness in the present, a link to a symbolic cultural bed. The final stage to evolve, literacy, is described as shared consciousness over time; literacy bypasses the limitations of person-to-person conversations.

Donald proposes this three-stage theory of consciousness in Darwinian terms: brains that exhibited more of these tendencies were selected because they were more adaptive than others. Language first evolved as an adaptation to help solve survival-related problems and became more complex in response to cultural challenges (such as language-mediated social interactions).

A PERSONAL VIEW OF DARWIN'S LEGACY IN PSYCHOLOGY

As stated earlier, contemporary psychology "grew up" with Darwin's ideas to such an extent that it is difficult to imagine an alternative. It should be noted, however, that psychology is in its infancy as a discipline and not all psychologists are Darwinists. No overall theory organizes or unifies psychology—although my personal bias is that Darwin's ideas come closest. So the future history of Darwin's influence on psychology will now be qualified.

As a professor of psychology in two large American universities for four decades, I've attempted to teach a Darwinian-biased version with mixed success. American culture, as reflected in its college students, is ill-prepared to struggle with the complexity and implications of Darwin's theories. *Darwin's Dangerous Idea*, the title of Dan Dennett's excellent book, is descriptive of the suspicious, accusatory bias instilled by some theists and others who either reject or deny humanity's biological underpinnings.[40] Many American college students rightly perceive that Darwinian thinking does change everything, especially, using Freud's term, their "naive self-love."

Another insight into the problems of teaching the "modern synthesis" (how genetics is related to natural selection) came from reading Ernst Mayr's *One Long Argument*.[41] The theory of natural selection is deceptively simple but modern synthesis is not. Natural history is not taught in most American schools, nor are any of the sciences taught particularly well. Mayr points out that one must integrate many disciplines to "understand" all that evolution entails: geology, paleontology, biology (systematics, animal behavior, and genetics), comparative anatomy, comparative physiology, comparative psychology, and behavioral biology. To this list we can add comparative genomics.

Perhaps the greatest hindrance to understanding cosmology and natural history is an inherent failure to comprehend vast time scales; our brains evolved to measure events during a human lifetime, and adding zeros to a given time frame doesn't adequately capture these changes. In the absence of education, and perhaps due to other cognitive limitations, much of modern science is taken on faith.

I have not formally studied the science of geological dating. I uncritically accept that the science that does so is sound and that estimates of the earth's age and the age of the universe reflect the best of human thinking. I do not have the sufficient mathematical background and am too lazy, or otherwise not competent, to "understand" fractals, chaos theory, and the processes attendant with a big-bang event. Again, I accept the methods employed and conclusions drawn by scientific "experts." Such acceptance is learned and is as much an emotional as a rationale one.

With respect to Darwin's theory, in many regards a large portion of American culture continues to reflect C. Lloyd Morgan's emotional response over a century ago. Where Darwin perceived "grandeur" in his view of life, others were consumed by uncertainty. In Dennett's words:

> Darwin explained a world of final causes and teleological laws with a principle that is, to be sure, mechanistic (and) utterly independent of "meaning or "purpose." It assumes a world that is absurd in the existentialist's sense of the term; not ludicrous, but pointless.[42]

In many respects Darwin's influence on American psychology is minimized by current textbooks that cater to unprepared students. When I recently wrote a general introductory psychology text, I was informed by the publisher that due to its evolutionary perspective, it would sell better in Canadian universities than in a broad swath of American universities.[43] Psychology professors select textbooks for their students, and not all psychology professors are exposed to Darwinian thinking in their specialty areas.

Indeed, I have many competent university colleagues in psychology and in other sciences, and in the humanities, who do not entertain a Darwinian world view. I try to remember this in attempting to teach psychology from an evolutionary perspective to 20-year-olds.

Conclusions

Psychology as a formal discipline was influenced from its inception by Darwinian thinking. William James's influential textbooks determined the broad scope of its subject matter, an adaptive "functionalism." Humans were part of the natural history of life on earth, and the human brain had a long evolutionary history which predisposed humans to behave in certain ways.

Darwin had more to say about brain, mind, and behavior than most academics are aware. The earliest psychologists were influenced more by Darwin than most academics (including many contemporary psychologists) are aware.

A psychology not influenced by Darwinian thinking is difficult to imagine, and that is his lasting legacy. ❧

Notes

1 Darwin, C. 1859. *On the Origin of Species by Means of Natural Selection*. London: John Murray, 499.

2 Fancher, R. 2009. "Scientific Cousins: The Relationship Between Charles Darwin and Francis Galton." *American Psychologist* 64 (2): 84–92.

3 Here, early Greek, Arab, Indian, and other writers' and philosophers' contributions will not be discussed.

4 Fancher, R. 1996. *Pioneers of Psychology*, 3rd ed. New York: Norton.

5 Galton, F. 1869. *Hereditary Genius*. Gloucester: Peter Smith, 1962.

6 Fancher, R. 1985. *The Intelligence Men: Makers of the IQ Controversy*. New York: Norton.

7 Galton, 1869.

8 Darwin, C. 1872. *The Expression of Emotions in Man and Animals*. London: John Murray.

9 Darwin, C. 1871. *The Descent of Man, and Selection in Relation to Sex*. London: John Murray.

10 Dewsbury, D. 2009. "Charles Darwin and Psychology at the Bicentennial and Sesquicentennial." *American Psychologist* 64 (2): 67–74.

11 Ibid.

12 Darwin, 1859.

13 Dawkins, R. 1976. *The Selfish Gene.* New York: Oxford University Press.

14 Burghardt, G. 2009. "Darwin's Legacy to Comparative Psychology and Ethology." *American Psychologist* 64 (2): 102–10.

15 Nesse, R. and P. Ellsworth. 2009. "Evolution, Emotion, and Emotional Disorders." *American Psychologist* 64 (2):129–39.

16 Darwin, 1871.

17 Buss, D. 2009. "The Great Struggles of Life: Darwin and the Emergence of Evolutionary Psychology." *American Psychologist* 64 (2): 140–48; Folkerts, D. 2012. Sexual Selection. In *Charles Darwin: A Celebration of His Life and Legacy.* Montgomery: NewSouth Books.

18 Darwin, 1871.

19 Jerison, H. 1955. "Brain to Body Ratios and the Evolution of Intelligence." *Science* 121: 447–9.

20 Jerison, H. 1973. *Evolution of the Brain and Intelligence.* New York: Academic Press; Kandel, E., J. Schwartz, and T. Jessell. 2000. *Principles of Neural Science*, 4th ed. New York: McGraw-Hill.

21 The author of this chapter is a fifth generation Tichenerian.

22 Fancher, 1996.

23 Wundt, W. 1897. *Ethics: An Investigation of the Facts and Laws of the Moral Life.* London: Sonnenschein, 153–4.

24 James, W. 1890. *Principles of Psychology.* New York: Henry Holt, 79.

25 Ibid.

26 Romanes, G. 1881. *Animal Intelligence.* London: Kegan Paul, Trench, Truber; Romanes, G. 1883. *Mental Evolution in Animals.* London: Kegan Paul, Trench, Truber; Morgan, C. 1882. "Animal Intelligence." *Nature* 26: 523–4; Morgan, C. 1894. *An Introduction to Comparative Psychology.* New York: Charles Scribners Sons; Washburn, M. 1908. *The Animal Mind: A Textbook of Comparative Psychology.* New York: Macmillan.

27 Morgan, C. 1903. *Introduction to Comparative Psychology*, 2nd ed. London: Walter Scott, 59.

28 McDougall, W. 1929. *Modern Materialism and Emergent Evolution.* New York: D. Van Nostrand and Company, 152.

29 Ibid.

30 Thorndike, E. 1909. "Darwin's Contribution to Psychology." *University of California Chronicle* 12: 65–80; Thorndike, E. 1911. *Animal Intelligence: Experimental Studies.* New York: Macmillan.

31 Thorndike, 1911.

32 Ibid.

33 Dennett, D. 1981. *Brainstorms.* Boston: MIT Press.

34 Pavlov, I. 1927. *Conditioned Reflexes: An Investigation of the Physiological Activity of the Cerebral Cortex.* Facsimile of the first edition. New York: Dover Publications, 8.

35 Miller, G. 2000. *The Mating Mind: How Sexual Choice Shaped the Evolution of Human Nature.* New York: Doubleday.

36 Cosmides, L. and J. Tooby. 2005. "Conceptual Foundations of Evolutionary Psychology." In *The Handbook of Evolutionary Psychology*. ed. D. Buss. New York: Wiley, 5–67. For recent overviews of relevant theory and research in evolutionary psychology, the interested reader may consult Cosmides and Tooby (2005).

37 Buss, 2005; Buss, D. 2008. *Evolutionary Psychology: The New Science of the Mind,* 3rd ed. Boston: Allyn and Bacon.

38 Schmidtke, K., J. Magnotti, A. Wright and J. Katz. 2012. The Evolution of Comparative Psychology. In *Charles Darwin: A Celebration of His Life and Legacy.* Montgomery: NewSouth Books.

39 Donald, M. 2001. *A Mind So Rare: The Evolution of Human Consciousness*. New York: W. W. Norton.

40 Dennett, D. 1995. *Darwin's Dangerous Idea*. New York: Simon and Schuster.

41 Mayr, E. 1991. *One Long Argument: Charles Darwin and the Genesis of Modern Evolutionary Thought*. Cambridge: Harvard University Press.

42 Dennett, 1981.

43 Barker, L. 2004. *Psychology*, 2nd ed. Upper Saddle River: Pearson Custom Publishing.

9

The Evolution of Comparative Psychology

Kelly A. Schmidtke, John F. Magnotti,
Anthony A. Wright, and Jeffrey S. Katz

Imagine yourself as a laboratory pigeon. Each day you are placed in a wooden box. Mounted on one wall is a computer monitor with a touch screen. That box is your world. In your world, two vertically aligned pictures and a white square appear on the screen (see Figure 9.1). At first, these pictures mean nothing to you and you do not do anything with them, but through experience these pictures become meaningful. You learn that when the two pictures are the same and you touch the bottom picture you are rewarded with food, but if you touch anywhere else you receive nothing. When the two pictures are different and you touch the white square, you are rewarded with food, but if you touch anywhere else you receive nothing. To survive you must eat, and so to survive in your world, you change your patterns of behavior from doing nothing with the pictures to reliably touching the bottom picture when the pictures are the same and touching the white box when the pictures are different. Changes in behavior patterns that help animals survive in their worlds are what we call learning. Comparative psychologists are interested in the systematic ways that behavior patterns change not only for pigeons but for all animals, including humans.

Arguably, the most important person in the creation of comparative psychology was Charles Darwin. As a child, Darwin developed a curiosity about animals and began learning the names of every animal around his English home while helping his brother with insect and bird experiments.

Sidetracking from this interest in nature, and perhaps following in his father's footsteps, he initially studied medicine but found surgical practice upsetting so he neglected his studies. Instead he spent his time with the Plinian society, a group of people who supported natural studies. After Darwin left medical school, his father then encouraged him to study theology; he lost interest in that, too. It seemed that nothing could pique Darwin's interest the way nature could. While continuing his core work in theology, Darwin also embraced his love of nature by taking courses such as botany and geology.

Finishing his degree, Darwin continued to expand his knowledge of nature on a five-year overseas tour. On this trip, he discovered a great number of new species, both extinct (as evidenced by their fossil remains) and living. Darwin noted the physical similarity between the extinct species and many of the living species he encountered. This realization caused Darwin to wonder by what mechanism new species replaced extinct ones. After more study, his answer would be evolution, which he described in *On the Origin of Species*, which was so popular that it sold out the first day it was available.[1]

In *Origin of the Species* Darwin focused on evolution with regard to physical continuity across species. Evolution described how existing organisms changed over time to survive in their worlds. Importantly, evolution is concerned only in maximizing survival value, not in advancing toward a specific end. A common misconception, known as *scala naturae*, is that evolution has led to the most advanced animal, the human.[2] *Scala naturae* can lead to inappropriate questions about the "best" or "most advanced" way to perform a specific task, without considering an animal's environment. For example, most water animals have gills to help extract oxygen; but to breathe outside water, land animals have lungs. Both gills and lungs are adaptive features, but which is most advantageous depends on your world (water or land).

In his second major book on evolution, *The Descent of Man, and Selection in Relation to Sex*, Darwin extended his thoughts on evolution from physical continuity to mental continuity. In this book he described how, like physical features, mental abilities can change to aid survival in one's world. Laying the foundations for comparative psychology, he stated that:

> All [animals] have the same senses, intuitions, and sensations—similar passions, affections, and emotions, even the more complex ones, such as jealousy, suspicion, emulation, gratitude, and magnanimity; they practise deceit and are revengeful; they are sometimes susceptible to ridicule, and even have a sense of humor; they feel wonder and curiosity; they possess the same faculties of imitation, attention, deliberation, choice, memory, imagination, the association of ideas, and reason, though in very different degrees.[3]

The key thing about Darwin's observations is that while he did note mental differences between species, those differences were of "degree" (quantitative) and not of "kind" (categorical). Put another way, of course humans can do some things other animals cannot do, but it is possible that everything a human can do, many other animals can do to a different extent. For example, while humans build skyscrapers, ants build tunneled mounds; while humans court love through poetry, birds court love through song; and while humans save extra money, nutcracker birds store extra food.

While Darwin made spectacular observations regarding mental continuity across species, systematic study needed to unfold to capture more precisely that continuity. One of the first attempts to capture animals' intellectual behaviors was undertaken by Darwin's friend George Romanes. Like Darwin, using an anecdotal method (which involved observing behavior and making inferences about the behavior but not experimentally testing the inferences), Romanes observed and eloquently wrote about the intellectual characteristics of animals' behavior. For example, after watching a cat open a latched door to escape from a room, he wrote,

> Cats in such cases have a very definite idea as to the mechanical properties of a door. . . First the animal must have observed that the door is opened by the [human] hand grasping the handle and moving the latch. Next, she must reason . . . If a hand can do it, why not a paw?[4]

Romanes colorfully recounted the intellectual feats of not only cats but also ants, dogs, primates, etc. He expressed that his anecdotal work would map "animal psychology for the purposes of subsequent synthesis," which acknowledges a major weakness of the anecdotal method.[4] This weakness is that the anecdotal method relies on a single observed event to capture the complicated nature of a learning process which unfolds across many events. Emphasizing this problem, Conway Lloyd Morgan stated that single observations, no matter how careful, would not be adequate for the interpretation of an animal's behavior.[5] Another problem with the anecdotal method is that it enables one to easily exaggerate species abilities, often by implicitly assuming the animals are solving the tasks in a human way (i.e., an anthropomorphic analogy). For example, perhaps the cat learned to open the latch not after a complicated process of reasoning about the mechanical properties of a door, but instead after a more simple process of random trial and error. Morgan captured his thoughts in his statement ("Morgan's Canon") that still guides the work of many comparative psychologists: "In no case may we interpret an action as the outcome of the exercise of a higher psychical faculty, if it can be interpreted as the outcome of the exercise of one which stands lower in the psychological scale."[5]

Around the time "Morgan's Canon" took hold of comparative psychology, Edward Lee Thorndike was already using a Darwinian framework in his comparative psychology research program. Thorndike formulated a general learning theory that describes how animals learn based on the consequences of their behaviors. Like Romanes, Thorndike also observed cats opening doors. Instead of escaping from a large room, however, Thorndike's cats escaped from small puzzle boxes in precisely determined ways (e.g., by pulling a string). Rather than hypothesizing about and discussing the cats' complicated reasoning about mechanical properties, Thorndike simply analyzed the amount of time it took the cat to escape from the box across repeated experiences of being trapped.

Thorndike observed that the cats seemingly did not apply any complicated, purposive strategy. Instead, the cats behaved randomly until they hit upon the right response and escaped (i.e., trial and error). Repeated experience in the box led the cats to repeat the same series of behaviors, such as spin around ➡ claw front ➡ claw left ➡ claw right ➡ pull string ➡ escape box. Behaviors that did not lead to escape were gradually "stamped out" (e.g., spinning and clawing) and behaviors that aided escape were "stamped in" (e.g., pulling a string). Eventually, only the behaviors that led to escape remained, and the time it took the cat to escape decreased across repeated experiences. Thorndike summarized his observations in his "law of effect," which states that the likelihood of a response's recurrence is generally governed by its consequence.[6]

Although the law of effect seems obvious to us now, Thorndike's work was critical in establishing its usefulness across various environments and species. For new environments, Thorndike introduced the cats to new boxes. He noted that in the same way escape times from the original box decreased with repeated trials, the time it took the cats to escape from the new boxes decreased. For new species, Thorndike directly compared the ability of fish, chickens, cats, dogs, and monkeys to perform an equivalent task, altering the puzzle boxes only to account for the physical differences between species (e.g., a fish's box must be in water). While some species escaped faster than others, the time it took each species to escape from their boxes decreased with repeated experience. The later application of the law of effect to various animal species shows the vast amount of continuity between species mental abilities, strengthening Darwin's original notion that mental differences among species were "of degree and not of kind." That is, the decrease in time and the ability to escape a puzzle box indicate species share the same mental faculties to solve the puzzles, whereas the fact that some species learn faster than others is a difference in degree.

Years later another highly regarded psychologist, Burrhus Frederic Skinner, extended Darwin's original ideas about evolution to impact the way we think about natural selection from the genetic level to behavior at the individual and group levels. He discussed how all levels of selection (genetic, individual, and group) function to increase the likelihood of features (genes, behaviors, or rituals) that aid an organism's survival and decrease the likelihood of features that are harmful. At the genetic level, genes that aid survival become increasingly likely in future generations; for instance, humans have a genetic code to produce lungs, rather than gills, to breathe. At the individual level, behaviors that aid survival—such as the pigeon touching the bottom picture when it is the same as the top picture (see Figure 9.1)—become increasingly likely in the future. Finally, at the cultural level, rituals that aid survival become increasingly likely in the future: greeting another person by shaking hands if you live in the United States or slightly bowing if you live in China to ensure future pleasant interactions.

Although researchers like Thorndike and Skinner explored the fundamental properties of learning and stressed continuity across species' mental abilities, much of their research focused on purposely simplified worlds and seemed to indicate that animals learned through trial-and-error processes. But the real world is more complex than such laboratory devices as boxes, and although learning may be simple in a simplified world, it does not follow that learning is simple in a complex world. In their defense, much of the work of early behavioral psychologists aimed at providing mechanistic explanations for behavior, in stark contrast to the anecdotal method.

Working in the Canary Islands, Wolfgang Köhler studied more complex learning by observing problem-solving behavior in chimpanzees. In a typical problem-solving task, a banana would be placed outside of the chimpanzee's cage, or inside the cage, but too high for easy access. The chimpanzees learned to use sticks to fetch the faraway bananas and stacked boxes to reach the bananas attached to the ceiling of their cages. Although these stick and box tools are less complicated than the tools that humans use, they are indeed tools, and so this research again supports Darwin's notion that differences between species are "of degree and not of kind."

Unlike Thorndike's results, Köhler's results suggest that the chimpanzees solved tasks purposively. Rather than a gradual stamping-in process of trial-and-error learning, Köhler's chimpanzees were able to solve the problem using insight. Just as humans solve brain teasers, Köhler's chimpanzees had no success until they figured out the answer (an "aha" moment), then they were extremely accurate. Problem-solving research continues to this day across species, and the methods different species use to solve problems shed light on how their mental abilities evolved.

Discoveries of intelligent behavior in animals continue to appear at an accelerated rate. Although examples of ape and monkey intelligence are not too surprising, even avians are casting off their "bird brain" label. Pigeons, for example, can navigate by landmarks, sun compass, magnetic compass, and infrasound sources. Galapagos finches and New Caledonian crows select and fashion tools (e.g., cactus spines twigs and leaves) to "fish" out insects and grubs in tree limb holes or cracks or under leaf detritus and carry their favorite tools around with them. Clark's nutcracker birds store thousands of pine seeds in hundreds of cache sites with sites varying yearly and often covered with snow by the time they are retrieved. These examples provide concrete support for shared, general processes across distally related species.

Although mental continuity is a hallmark of Darwin's theory, there is an important corollary: animals should also develop specific abilities based on the worlds in which they live (i.e., ecology-specific abilities). While basic mental abilities may be the same across species, organisms should have differences based on when and where they evolved. For example, bats hunt for food during the night so they have developed a specific ability to use echolocation, a series of clicking noises that echo back to them and allow them to judge the distance between themselves and their prey, (i.e., to locate food). Humans, on the other hand, hunt (or shop) for food during the day and can rely on vision to find nourishment. While there are cases of humans successfully using echolocation, the specific ability to use echolocation is less precise in humans, again supporting Darwin's notion that differences between species are "of degree and not of kind."

The last 30 years have seen an increase in this ecological research, exemplified by the work of Sara Shettleworth and her ecological program, which emphasizes how the world influences the development of particular mental abilities.[7] Shettleworth cautions researchers to avoid an anthropocentric (human-centered) research focus, which does not consider species' ecologies, asking, for example: Can nonhuman animals escape from boxes like humans, or, Can nonhuman animals remember where food is located like humans? Instead, she suggests that researchers study closely and distally related species to better understand the impact of specific ecologies. The researcher is then challenged with identifying and manipulating the critical parameters, such as the delay between seeing a clue that indicates where food is located and a later opportunity to obtain it in a memory task. By systematically manipulating the critical parameters patterns often emerge that allow for more appropriate comparisons across species. For example, while all species may find it more difficult to remember something as time passes (a qualitative similarity), quantitative differences may emerge in how quickly a given species' performance declines over time.[7]

In summary, to better understand the genetic and ecological contributions to mental abilities one should compare closely and distally related species with a conceptually equivalent task. Debora Olson did just that to compare memory for food storing birds: she directly compared the spatial memory of Clark's nutcrackers to a closely related species, the scrub jay, and a more distally related species, the pigeon, in an equivalent spatial-memory task.[8] If spatial memory was a genetically determined ability, then one would expect the closely related birds to perform more similarly than the distally related birds.

In her task, each bird was placed in a box with four buttons in the front and one button in the rear. In the box, birds touched buttons to receive rewards. In a trial, the rear button illuminated first, which the bird touched to extinguish. Then one of the four buttons at the front (a sample location) illuminated, which the bird also touched to extinguish. Then the rear button illuminated again, and after the bird extinguished it, a waiting period began (e.g., a 1-second delay). After the delay, two of the front buttons illuminated. Now the bird had to decide which button to touch. If the bird chose to touch the button that it had not previously touched, then it was rewarded with food; otherwise, it was not. A correct choice would indicate the bird's accurate memory for a spatial location. On the next trial, the period of delay between touching the rear button and the choice buttons illuminating was determined by how the bird performed on the previous trial. If the bird responded correctly on the previous trial, then the delay lengthened; however, if the bird had responded incorrectly, the delay decreased. Progressing and performing accurately at longer delays indicated better memory for spatial locations over longer intervals.

Differences of degree but not of kind were found between the species. All birds tolerated delays over 0 seconds (the front buttons are lit immediately after the rear light extinguishes), indicating all species were able to perform the spatial memory task. However, differences emerged in the delay length that each species could tolerate. Pigeons tolerated the shortest delays, .5 to 25 seconds, followed by the jays, 7 to 44 seconds, and finally the nutcrackers, 50 to 80 seconds. Later experiments revealed that species differences between caching birds, like nutcrackers, and non-caching birds, like pigeons and jays, emerged only in spatial memory tests. When caching and non-caching birds' performances are compared without the spatial component, their ability to remember is very similar.[9] These results show that nutcrackers have a sophisticated species-specific ability to remember spatial locations that is quantitatively better but not qualitatively different than pigeons and jays.

To summarize, the ecological program's goal is to directly compare the performance of closely and distally related species in equivalent tasks to learn about genetic and ecological contributions to mental abilities. If closely related species that live in different ecological niches perform similarly, it is assumed that those mental abilities are more influenced by genetic than ecologic contributions. Alternatively, when two distally related species that live in the same environment perform tasks similarly, it is assumed that those mental abilities are more greatly influenced by ecological than genetic contributions. In short, when genetic differences are large, the environmental influence is small, but when the genetic differences are small, the environmental influence is large.

Currently, comparative psychologists study a diverse array of topics, from simple processes, such as sensations, to more complex processes, such as episodic memory, language use, imitation, and reasoning, just like Darwin outlined more than 100 years ago. These research programs ask, and provide initial answers to, important comparative questions, but many do not use direct comparisons in equivalent tasks, like Thorndike's puzzle box or Olson's spatial memory research programs. As one contemporary example, research in our laboratories has focused on another area of comparative cognition, the development and deployment of abstract concepts in humans, monkeys, and pigeons. Our findings show the extent to which abstract-concept learning is a general process conserved across diverse species.

Abstract concepts are a type of knowledge that transcends the specific items used to represent them. For example, when you were young you learned item-specific math problems, like 2 times 2. Currently, there is no need for you to do any math to confidently answer that problem because you have memorized the answer item specifically and know it by rote (the answer is obviously 4). However, if you are presented with a math problem that you have not memorized, like 456 times 23, you could answer it without having past experience with it because you know abstract math concepts (in this case, multiplication rules). If you do not believe it, get out a pad of paper and do the equation and then check your answer with a calculator. Our laboratory has chosen to focus on a more fundamental abstract concept, same/different. This concept is abstract in the same way the math concept is abstract. To prove it to yourself, hold any two objects in front of you and state if they are the same or different. The objects (such as pens or pencils) themselves are not same or different, but the relation between any two objects can be the same (two pens) or different (a pen and a pencil).

In our research program we directly compare different species in an equivalent task. In this task, two pictures and a white square appear on a computer screen. If the two pictures are the same, the correct response is to touch the bottom picture; if the two pictures are different, the correct response is to touch the white square (see Figure 9.1).

After learning the task with eight specific pictures, novel pictures are introduced. If our subjects can perform the task accurately with novel pictures, we can say that the subjects have the abstract same/different concept. However, if the subjects perform less accurately with the novel pictures than the trained pictures, then we cannot say they have fully learned the abstract same/different concept. Instead, we can say they likely learned item-specific rules to perform the same/different task with specific pictures. (This is just like solving 2 times 2 but with pictures. Consider taking a math exam on multiplication. If you solved all the problems correctly and received a 100, you fully learned the abstract math concept, but if you answered

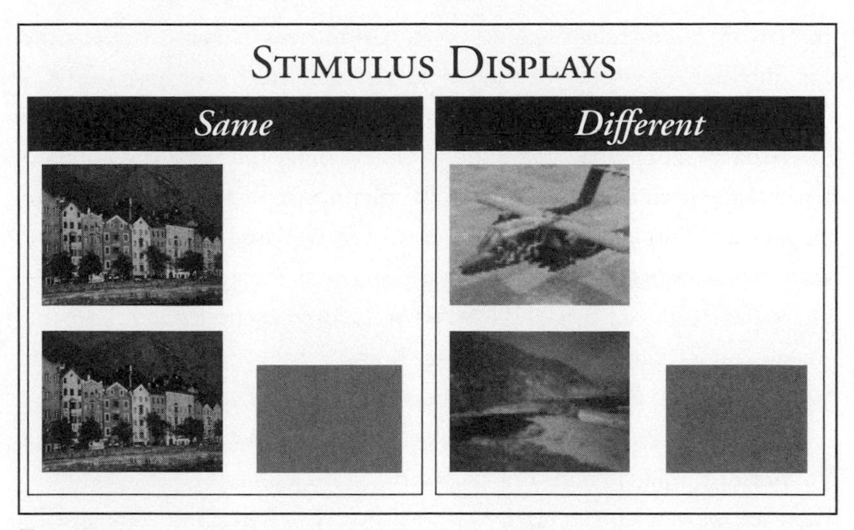

Figure 9.1

The stimulus displays used to train the humans, rhesus monkeys, capuchin monkeys, and pigeons. A touch to the bottom picture was correct on same trials. A touch to the gray rectangle was correct on different trials. The actual stimulus displays did not have labels, had black backgrounds, and the gray area to the right was white.

only 70 or 80 percent correctly, you have only partially learned that math concept.) But this is not where our experiments end. Instead the training set size is increased to include more pictures, and subjects are tested with novel pictures after learning the task with progressively larger training set sizes ranging from 16, 32, 64, 128, 256, 512, and 1,024 pictures.

Let us start with humans' performance in our same/different task. Humans can be difficult to test because they enter our laboratory with so much previous experience related to the same/different concept. And while we assume that all of our college-age participants have an abstract same/different concept, this assumption is no better than using Romanes' anecdotal method. So instead of assuming that the humans know the same/different concept, we invite them into our laboratory room to experience the same/different task on a touch screen computer. To avoid the humans easily applying a concept they have already mastered, we tell them they will be completing a "touch screen proficiency task." During the task the humans are rewarded with points on the computer screen for correct same and different responses. After learning the task, nearly all humans were able to perform the same/different task accurately with novel items after experience with only eight specific pictures. Those humans who failed to perform the task accurately with novel pictures reported using elaborate strategies to complete the task, such as strategies related to how pretty they judged the pictures or how firmly they touched the screen, indicating they had over-thought the task and learned an incorrect set of rules. That said, humans can perform the same/different task after experiencing only eight specific pictures. What about monkeys and pigeons?

We have looked at capuchin[10] (a new world monkey) and rhesus[11] (an old world monkey) monkeys' performance in our same/different task. These are primate species like humans, but we made some changes for nonhumans. Monkeys were tested unrestrained in large metal boxes instead of the laboratory room. Also, we did not use the "touch screen proficiency task" guise to hide the true purpose of the task. Because monkeys are not interested in points, they were rewarded with juice and banana pellets for correct same and different responses. Lastly, the monkeys had to be trained to pay attention to the top picture by requiring them to touch the top picture before the bottom picture and white rectangle appeared. These task modifications (environment, previous knowledge, reinforcement, and attention) did not preclude the monkeys from learning the same/different task with eight specific pictures. However, after learning the task with eight specific pictures, the monkeys did not perform the task accurately with novel pictures. If the experiment stopped here, then we could have concluded a species difference in kind: humans but not monkeys have the mental ability to use the same/different concept. But, as previously mentioned, the experiment does not stop here. Instead, the training set size is increased to include more pictures and subjects are tested with novel pictures after learning the task with larger set sizes. Indeed, monkeys showed better transfer with larger training set sizes, and after training with 128 specific pictures, the monkeys performed as accurately with novel pictures as with their training pictures. This emphasizes the difference between human and monkey mental abilities to use the same/

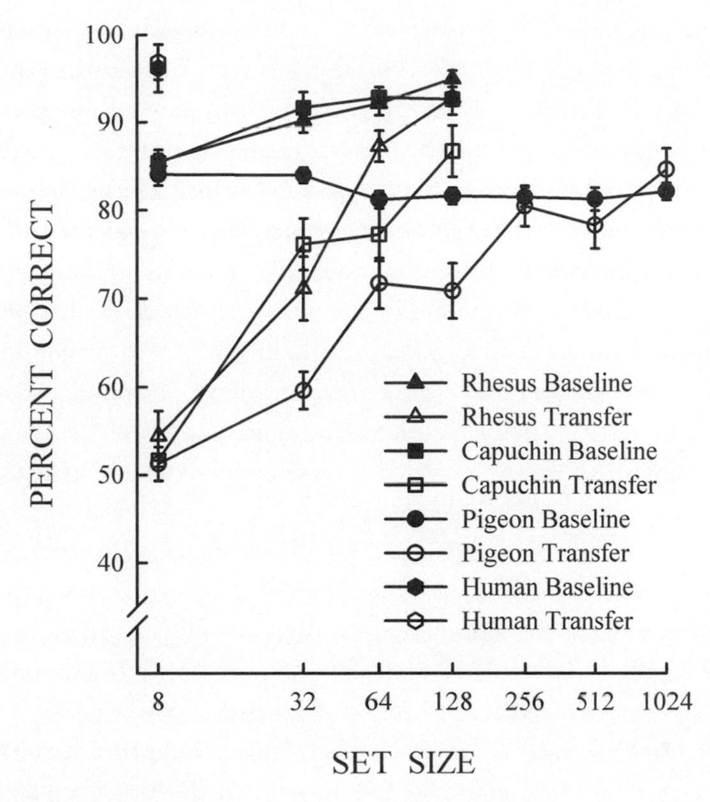

FIGURE 9.2

Mean performance and standard errors for baseline and transfer performance at each set size for humans, rhesus monkeys, capuchin monkeys, and pigeons. The set size where baseline and transfer performance are equivalent represents full abstract-concept learning: 8 for humans, 128 for monkeys, 256 for pigeons.

different concept is of "degree and not of kind."

We have also looked at pigeons' performance in our same/different task,[12] a species distally related to humans by about 300 million years. Again, species differences unrelated to the same/different concept were evident. Pigeons are small and weak compared to humans and monkeys, so they were tested in small wooden boxes. Also, unlike humans and monkeys, pigeons have no fingers; therefore, they cannot touch the screen in the same manner. So instead of requiring a finger touch, the pigeons touched the pictures with their beaks. In addition, because pigeons are not interested in points, banana pellets, or juice, they were rewarded with grain for correct same and different responses. Finally, it was necessary to require the pigeons (like the monkeys) to touch the top picture before the bottom picture and the white square appeared. Just like the monkeys, we found that the pigeons failed to perform accurately with novel pictures after learning the task with eight pictures, but when we progressively increased the training set size, we also found better transfer to novel pictures. For bird-brained pigeons, we had to enlarge the training set size to include 256 specific pictures to show full abstract concept learning. So, humans show full abstract concept learning at 8 pictures, monkeys at 128 pictures, and pigeons at 256 pictures in our task. This represents a difference in degree but not in kind across these different species.

The differences in degree between humans, monkeys, and pigeons in the same/different task are displayed in Figure 9.2. Figure 9.2 shows that the transfer performance increased across set size until full concept learning occurred (humans = 8, monkeys = 128, and pigeons = 256). Importantly, Figure 9.2 shows no difference in kind, as all species eventually performed as accurately with training and novel pictures, demonstrating that they had fully learned the abstract same/different concept. In the future, we hope to expand our research program to investigate abstract-concept learning with a greater number of species.

In conclusion, our research (like that of many others) continues to support Darwin's notion that animal differences are of degree but not of kind. ☙

NOTES

1 Darwin, C. 1859. *On the Origin of Species by Means of Natural Selection, or the Preservation of Favoured Races in the Struggle for Life*, 1st ed. London: John Murray.

2 Hodos, W. and C. Campbell. 1969. "Scala Naturae: Why There is No Theory in Comparative Psychology." *Psychological Review* 76: 33–50.

3 Darwin, C. 1871. *The Descent of Man, and Selection in Relation to Sex*. New York: D. Appleton and Company.

4 Romanes, G. 1904. *Animal Intelligence*, 8th ed. London: Kegan Paul, Trench.

5 Morgan, C. 1894. *An Introduction to Comparative Psychology*. London: W. Scott.

6 Thorndike, E. 1911. *Animal Intelligence*. New York: Macmillan.

7 Shettleworth, S. 1993. "Where is the Comparison in Comparative Cognition? Alternative Research Programs." *Psychological Science* 4: 179–84.

8 Olson, D. 1991. "Spatial Memory in Clark's Nutcrackers, Scrub Jays and Pigeons." *Journal of Experimental Psychology: Animal Behavior Processes* 17: 363–76.

9 Olson, D., A. Kamil, R. Balda and P. Nims. 1995. "Performance of Four Seed-Caching Corvid Species in Operant Tests of Non-Spatial and Spatial Memory." *Journal of Comparative Psychology* 109: 173–81.

10 Wright, A., J. Rivera, J. Katz and J. Bachevalier. 2003. "Abstract-Concept Learning and List-Memory Processing by Capuchin and Rhesus Monkeys." *Journal of Experimental Psychology: Animal Behavior Processes* 29: 184–98.

11 Katz, J., A. Wright and J. Bachevalier. 2002. "Mechanisms of *Same/Different* Abstract-Concept Learning by Rhesus Monkeys (*Macaca mulatta*)." *Journal of Experimental Psychology: Animal Behavior Processes* 28: 358–68.

12 Katz, J., and A. Wright. 2006. "*Same/Different* Abstract-Concept Learning by Pigeons." *Journal of Experimental Psychology: Animal Behavior Processes* 32: 80–86.

10

Evolution and Embryology

How Developmental Biology Plays Into Evolutionary Theory

Kenneth M. Halanych

In the latter half of the 1800s, most people would not have distinguished between the fields of embryology and developmental biology. Today, we recognize embryology as a sub-discipline of the more holistic field of developmental biology. Perhaps the easiest way to think of the difference between theses two fields is that organisms usually continue to grow or develop after embryogenesis (e.g., a baby human continues to develop after birth). Like evolution, embryology was undergoing rapid development in the 1800s. In fact, several fields of biology were experiencing such growth because, for the first time, scientists were undertaking systematic studies of animal and plant anatomy. Additionally, expansion of national empires in the 1700–1800s allowed a flood of unique and unusual samples from around the globe to reach these comparative anatomists. This confluence forced scientists to address some very basic questions: Why do things look the same? Why do they look different? And how did they get that way?

The contributions of Charles Darwin to the development of modern evolutionary theory, the unifying theory of biology, cannot be overstated. Darwinian ideas have influenced all aspects of biological thought, as well as societal development, astrophysics, and the development of vaccines and pharmaceuticals. Darwin's *Origin of Species* was a brilliant and stunningly thorough capstone in a growing 19th-century movement to understand and explain organismal variation and diversity within the context of natural laws.[1] However, Darwin did not do this in a vacuum. His ideas and insights on evolution were cultivated from other scholars. In this chapter, I will focus on the important role that developmental biology played in strengthening Darwin's ideas. Then my attention will be turned to the synergy, or lack thereof, between developmental biology and evolutionary theory since Darwin's time.

WHAT *Origin of Species* COVERED

Before launching into the role of embryology and developmental biology in Darwin's ideas, perspective is needed as to what *Origin of Species* did and did not encompass. This work is often credited with establishing five major concepts in biology:

1. ***Natural selection.*** The idea that variation within a group of individuals allows some to fare better, and others worse, in the acquisition of limited resources, including access to mates. With a sufficiently limited resource, even small variations in obtaining those resources allow some individuals to survive, or mate, while others do not.

2. ***Population change.*** Darwin recognized that populations of individuals are not static, or immutable. Thus, within a population, traits favored or disfavored by natural selection would not have the same prevalence over time.

3. ***Change over time.*** He also recognized that traits were heritable between parents and offspring. If prevalence of traits can change from generation to generation, then over multiple generations the amount of change can be significant. Thus, there can be considerable change over long time scales.

4. ***Gradualism.*** Darwin recognized that in many cases change would occur slowly and accumulate over time; that is, changes occur in a gradual manner. We now know that there can also be very abrupt changes in a small number of generations, for example, by punctuated equilibrium. (Darwin may have realized that given the intellectual climate at the time, the idea of rapid change may have been too radical to make his theory "acceptable." Some of his colleagues tried to convince him to soften his views on the pace of evolution.)

5. ***Common descent.*** The idea that two distinct forms or species share, through a line of descent, a common ancestor is central to modern evolutionary theory. Darwin recognized that through a combination of processes—including inheritance, natural selection, and reproductive isolation—that a once-cohesive population could form two daughter lineages that differ due to the effects of selection.

Darwin knew that his theory and ideas were a major step forward and would be controversial to some. Ingeniously, he began *Origin of Species* with detailed examples drawn from domestic animals. Given that it was the 19th century, the presence of domestic breeds and breeding practices to promote desired characters were much more commonplace than today. He also realized that there were aspects to evolutionary theory that were not yet understood. Two major aspects of modern evolutionary theory that were integrated after Darwin include the following:

1. ***Inheritance.*** Darwin recognized that offspring obtained traits from their parents, but he freely acknowledged that, at the time, "the laws governing inheritance . . .[were] . . . quite unknown."[2] That is, he did not know about genetic mechanisms that allow traits to be passed from parents to young.

2. ***Sources of variation.*** Darwin's theory did not fully or accurately explain how new variants were produced. We now understand mutations and rearrangements to DNA, the genetic material of life, are the ultimate source of variation.

Integrating mechanisms of inheritance with Darwin's theory of descent with modification via natural selection took place during the 1930s and 1940s and was called the "modern synthesis" of evolutionary theory. (Interestingly, some 70 to 80 years later it is still typically referred to as the modern or new synthesis). Lacking a mechanism of inheritance, Darwin had to turn elsewhere to explain his ideas about descent from a common ancestor. The emerging field of embryology, especially strong in Germany at the time, provided crucial examples of how organismal morphologies could diverge from a common starting point. Using this information, Darwin naturally focused on embryological and developmental concepts and examples to "complete" his ideas on common descent.

ORIGINS OF EVOLUTIONARY THEORY

If Darwin had not written his famous book, we would still have a solid understanding of evolutionary processes today. However, certain ideas, arguments, and examples used over the 20th century would have been different. Anyone who has read Malcolm Gladwell's book *Outliers* already knows that the right person being in the right place at the right time can lead to extraordinary results.[3] So it was with Darwin and *Origin of Species:* this book was the culmination of several influences and opportunities. Basic evolutionary ideas started well before Charles Darwin. In fact, his grandfather Erasmus Darwin (1731–1802) had published on "evolutionary" ideas in 1794, but Erasmus died before Charles was born and thus his influence was only indirect with Charles reading the senior Darwin's work.[4] A contemporary of Erasmus Darwin, Jean-Baptiste Lamarck (1744–1829), is better known than the senior Darwin for his evolutionary ideas, usually called the theory of transformation. Lamarck recognized that organisms pass on traits via inheritance and that organismal lineages change. However, he was off the mark on how variation was generated: he supposed that heritable changes were acquired over an individual's lifetime as a result of adapting to one's environment. A more direct influence on Darwin was Robert Edmond Grant (1793–1874), who instructed Darwin in the mid 1820s and began publicly voicing his support for Lamarckian ideas at the same time. Importantly, Grant was a materialist, not a theist. That is, he sought to root his explanations of nature in natural causes, not in the supernatural.

Darwin had formulated the basis of his theory by 1838, shortly after his amazing voyage on HMS *Beagle* (1831–36), which exposed him to numerous plants and animals. However, his ideas would not be published for another 20 years. In the intervening time, the logic and arguments underlying his thoughts would mature. During those 20 years, Darwin produced several other meaningful biological works. Moreover, the time allowed him to ruminate on other scholarly works and interact with other scientifically influential individuals who directly, or indirectly, helped him flesh out details for *Origin of Species*. Remember, Darwin was in the United Kingdom at the height of its Golden Age and had relatively easy access to the world's greatest minds in London and central Europe. Other individuals commonly attributed to influencing Darwin include Charles Lyell (1797–1875), who contributed to Darwin's thoughts on gradualism, and Thomas Malthus (1766–1834), whose economic ideas underpin the notion of selection due to limited resources. Others certainly caused Darwin to refine and fortify his arguments. For example, Richard Owen (1804–92), a comparative anatomist and paleontologist, worked on samples Darwin brought back on HMS *Beagle*.[5] Owen was a stalwart theist and supported the idea of archetypes (i.e., model forms). Owen later became a renowned critic of *Origin of Species*. However, he also argued with Robert Edmond Grant, Joseph Dalton Hooker (1817–1911), and Thomas Henry Huxley (1825–95) and earned a reputation for being underhanded. In many ways, Owen was not a nice man, and Darwin came to loath him late in life.

The fact that Darwin was cultivating evolutionary ideas was no secret. He regularly and openly discussed ideas with several prominent naturalists of the time, including botanists Hooker and Asa Gray (1810–88). Thus, when, in 1844, *Vestiges of the Natural History of Creation*,[6] a book focusing on evolutionary ideas and the transmutation of species, was anonymously published, some speculated Darwin had penned it.[7] After several editions, Robert Chambers (1802–71) was revealed as the author. Notably, *Vestiges* overlapped with several of Darwin's insights into evolution, but, despite being initially popular, *Vestiges* was openly criticized on all fronts. Darwin, like other contemporaries, was disappointed in the work because of numerous mistakes and superficial, as well as unsupported, speculations presented in its pages.[8]

As is widely known, Alfred Russel Wallace (1823–1913), who was at the time in the Malay Archipelago, proved to be the catalyst for the publication of *Origin of Species*.[9] Wallace had many of the same intellectual influences as Darwin; in particular, Wallace understood the implications of Malthusian economics for natural populations and had considerable experience in natural systems, especially archipelagoes. In 1858, Wallace sent Darwin his essay entitled "On the Tendency of Varieties to Depart Indefinitely from the Original Type," asking him to pass it to Lyell. Just prior to this, Wallace had briefly corresponded with Darwin, and Lyell knew of Wallace's earlier work. Darwin was impressed with Wallace's efforts and passed the paper to Lyell, who showed it to Hooker. All three recognized that Wallace had, more or less, come up with a theory identical to Darwin's natural selection. At the time Darwin was faced with several personal issues, including the death of his son, and Lyell and Hooker took it upon themselves to unveil the Darwin-Wallace evolutionary theory at the Linnean Society in London. Neither Darwin nor Wallace was present at the meeting; although the papers were jointly presented, Lyell and Hooker made sure that the priority of Darwin's ideas was clearly established.

The Role of Embryology in Darwin's Ideas

To Darwin, embryological data unequivocally supported his theory. He devoted most of Chapter 13 in *Origin of Species* to embryological evidence. There he states, "... the leading facts in embryology ... are second in importance to none in natural history ..."[10] Even after publication, he continued to espouse the importance of developmental data. In 1860, he wrote to his American botanical colleague, Asa Gray, "Embryology is to me by far the strongest class of facts in favor of change of forms."[11] Because of the comparative underpinnings of embryology at the time, his embryological discussions were coupled with the discussion of how organisms should be classified. As mentioned above, to Darwin, embryological evidence was critical to understanding common descent. This was particularly true while mechanisms of inheritance remained unknown.

Some embryological ideas that significantly influenced Darwin came from Karl Ernst von Baer (1792–1876), often credited as the father of embryology.[12] Von Baer had focused on relationships of embryological forms and radically broke from previous theories, including those of Lamarck. Von Baer's ideas have been canonized (for lack of a better word) into "von Baer's laws of embryology," which include the following:

1. General characters of a taxon develop in the embryo before the more specialized, or specific, characters.

2. Spatial relationships between general structures are established before relationships of more specific structures.

3. As an embryo develops it does not converge upon other definite forms,

but instead the embryonic form diverges from other embryonic forms.

4. Embryos of more derived animal forms never resemble the adult of more basal (less derived) forms; they only resemble the embryo of those forms.

Von Baer's observations were totally incompatible with Lamarckian theory. In the 1800s, many scientific scholars were dependent upon, and aligned with, various religious establishments. This was perhaps more true in Darwin's England than von Baer's Germany, where such alignments were based on a variety of pragmatic reasons such as financial support, access to libraries, and politics. Thus, finding concrete evidence that refuted the ideas of Lamarck, a transformationist, had significant repercussions for maintaining the role of the "archetype" in a theistic worldview. In essence, contemporary colleagues viewed von Baer's theories as a refutation of evolutionary theories.

Von Baer and Darwin apparently did not have much direct exchange of ideas, and von Baer's discoveries in embryology appear to have been conveyed by other sources. Henri Milne-Edwards (1800–85), with whom Darwin corresponded, was one of those responsible for importing von Baer's ideas to England. Another influence was likely American scientist Louis Agassiz (1807–73). In a few places in *Origin of Species*, Darwin discussed embryological data in the context of Agassiz. Darwin states, "The embryo, also, of distinct animals within the same class are often strikingly similar: a better proof of this cannot be given, than a circumstance mentioned by Agassiz, namely, that having forgotten to ticket the embryo of some vertebrate animal, he cannot now tell whether it be that of a mammal, bird, or reptile."[13] This anecdote was published earlier by von Baer and misattributed by Darwin in the first edition to Agassiz. By the third edition of *Origin of Species*, Darwin had corrected his mistake.

However, perhaps the main contribution von Baer provided Darwin was the idea of common descent. The only figure in *Origin of Species* is a diagram illustrating the divergence of taxa. Darwin had clarified and developed this figure from a sketch he had made around 1837 in one of his notebooks (Figure 10.1). The earlier illustration is presumably based on Martin Barry's (1802–55) drawing in which he attempted to capture the

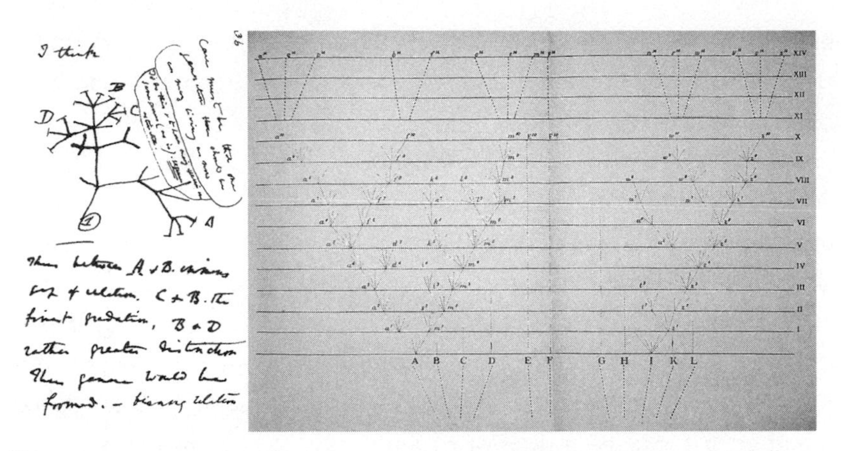

FIGURE 10.1
A) Darwin's sketch of a tree showing relatedness from the "B" notebook in 1937. B) The more mature version of the tree shown in Fig 1A. This tree, which depicts common descent, was the only illustration in ***Origin of Species.***

idea of common descent implied by von Baer's laws. Tracing the origin of a tree figure back further becomes more difficult (at least for me), but there is some suggestion that von Baer had utilized such diagrams to illustrate how embryos diverged from each other. Ironically, while trying to argue against the mutability of species forms, von Baer provided a key piece to the evolutionary puzzle. His third law listed above seems to necessitate a common ancestral form. In the case of Darwin, divergence from a common form was caused by natural selection.

HAECKEL: AN EMBRYOLOGIST AFTER PUBLICATION

Darwin thought embryology was the strongest type of data supporting his evolutionary ideas. As expected, immediately after publication, the scientific world began to take sides and to "arm" itself with the best evidence for and against Darwin's ideas. Opposing Darwin were Richard Owen, Louis Agassiz, and even Ernst von Baer, whose ideas were claimed by Darwin as a source of support. In his corner, Darwin had Thomas Henry Huxley (1825–95), Asa Gray, Joseph Dalton Hooker, and a young German naturalist and embryologist, Ernst Henrich Philipp August Haeckel

(1834–1919).[14] In the period after publication of *Origin of Species*, Haeckel was arguably the best-known embryologist and naturalist—for positive and negative reasons.

Haeckel, a German, was about 25 years old when *Origin of Species* was published. Prior to Darwin's publication, Haeckel had spent time on the Italian coast looking at a group of single-celled silica-coated organisms called radiolarians (Figure 10.2). Haeckel had noticed the variation and form in these stunning organisms. This experience allowed Haeckel, when he read *Origin of Species*, to immediately place Darwinian ideas in an organismal context based on personal experience. As a result, Haeckel was an ardent supporter of Darwin. By any standard, Haeckel's body of scientific work is amazing, and he coined several scientific terms (e.g., ecology and phylogeny) commonly used today.

The artistic talent of Haeckel was most impressive, and reproductions of his organismal lithographs are still widely circulated today. However, when inaccuracies were discovered in his drawings comparing embryonic forms, Haeckel was accused of academic dishonesty for intentionally modifying the organisms' appearance so they would fit his theories. In retrospect, Haeckel failed to acknowledge the "inaccuracies" in a timely and public manner. However, modern attacks on Haeckel for being a fraud are perhaps out of place. In the mid- to late 1800s, mechanisms for disseminating drawings were limited and expensive. Thus, drawings were typically made to accentuate certain features. This practice, to a lesser extent, is still in common use today. Consider that most modern organismal field guides, (e.g., a bird guide) use drawings instead of photographs because drawings allow distinctive features and markings to be highlighted and emphasized.

In addition to artistic ability and inaccurate images, Haeckel is known for coining the term "ontogeny recapitulates phylogeny." Translated into everyday terms, the developmental series that an organism goes through from zygote to adult summarizes, or replays, its evolutionary history by passing through ancestral forms. In other words, he argued that stages of a mammalian embryo pass through fish, amphibian, reptile, and earlier mammal stages. Ironically, this notion, which was advocated by a leading Darwin supporter, is incredibly non-Darwinian because it implies a linear progression of change (similar to Lamarck), rebutting the notion that species diverge from one another. Furthermore, according to Robert J. Richards, Haeckel was the individual perhaps most responsible for popularizing Darwin's theories.[15] Von Baer had argued against Lamarckian theory because his observations were not consistent with transmutations (changes in form within an organismal lineage), and in doing so clearly took a stand

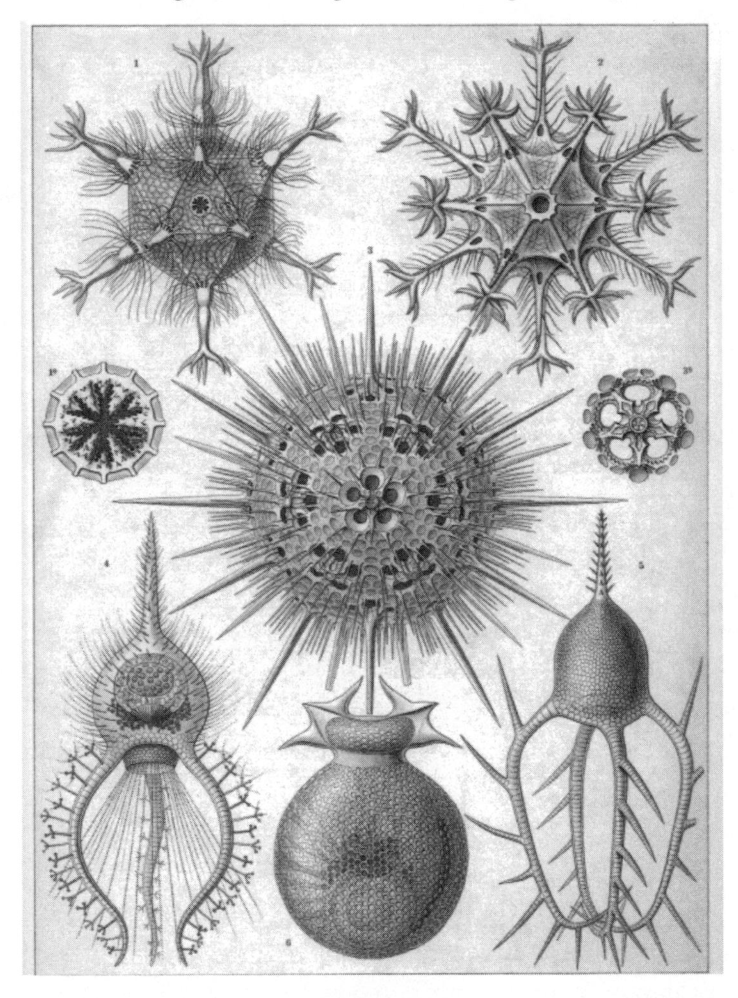

FIGURE 10.2

Reproduction of lithograph from Ernst Haeckle work on Radiaolarians. Reproduction from first edition of E. Hackel. 1904. Kunstformen de Natur, Leipzig and Vienna, Bibliographes Institut.

against ideas that a species relived its evolutionary past during the process of development. So in an odd juxtaposition, von Baer, whose embryological observations provided compelling evidence for modern evolutionary theory, was anti-Darwinian, but Haeckel, who was one of Darwin's most steadfast supporters, believed in an idea that was clearly non-Darwinian, a linear progression of forms. Thus, Haeckel effectively turned back the clock on scientific understanding in embryology. Richards also argues that Darwin held to a softer form of recapitulation ideas. However, if Darwin did hold such ideas, they took a back seat to his view that animals diversified from a common ancestor.

Interestingly, Haeckel and Darwin began corresponding after Haeckel sent Darwin his first monograph on radiolarians in the early 1860s, which incorporated the new theory laid out in *Origin of Species*. Haeckel even visited Darwin late in Darwin's life but the German-English language barrier apparently limited their conversation. They clearly influenced each other's later works, especially on issues relating to the descent of man. Both were aware of the implications on human origins, which Darwinian evolution raised. Namely, it challenged a theistic view in understanding nature. Whereas Darwin was subtle about this, Haeckel was forthright and, by some accounts, even antagonistic, about the fact that our understanding of nature should not be based on theistic underpinnings. Richards discusses Haeckel's role in polarizing Darwin's opponents, most of whom were theists.

Huxley also corresponded directly with Haeckel, and urged him to tone down some of his works before they were translated from German to English so as not to further instigate Darwin's opponents. To put this in perspective, Huxley's tenacity in defending Darwin's ideas earned him the nickname "Darwin's Bulldog." Thus, for Huxley of all people, asking Haeckel to be more restrained gives a sense of how divisive Haeckel could be. The boldness and acid nature of Haeckel's statements about survival, the role of death, and the finality of life may have been motivated by personal tragedy, including the death of his first wife at a critical juncture in his professional career. However, Haeckel appears to have accurately understood the philosophical difference between science and religion. Unfortunately, 150 years after publication of *Origin of Species*, large segments of societies across the globe fail to understand that science seeks explanation of the natural world based on natural causes and that religion has had a different role in society. In many ways, Haeckel was the Richard Dawkins of the day and would have enjoyed reading Dawkins's 2006 book *The God Delusion*.[16]

Before we leave Haeckel on a negative note, he deserves credit for popularizing and stylizing tree diagrams to illustrate common descent. This basic form of representation is the most common form used in current evolutionary literature. Throughout Haeckel's work there is interesting variation in his artistically drawn trees. For example, some of his early trees are rich with branches to the point of appearing shrub like. However, his stem tree of man's ancestry in *Anthropogenie; oder, Entwickelungsgeschichte des Menschen* (1874) emphasizes one main lineage with a limited number of side branches. At some level, this drawing is based on recapitulationist ideas.[17]

The Divorce

During Darwin's lifetime and shortly thereafter, embryology played a vital role in the development of modern evolutionary theory. However, after the beginning of the 20th century, the close link between evolutionary theory and embryology began to fade. There were likely several reasons for this change—personalities involved were not as dynamic and technical improvements were not adding major new insights, making the main issues matters of interpretation of existing data—but the overarching reason was the rediscovery of Gregor Johann Mendel's (1822–84) genetic studies on inheritance in pea plants. By Darwin's own admission, mechanisms of inheritance were an important part of a holistic understanding of evolutionary processes. Once Mendel's work was rediscovered, scientists set to work integrating the mechanism of inheritance into a Darwinian framework. But, first, geneticists had to get on the same page as they were arguing between the importance of phenotype or genotype during the first part of the 20th century. By the 1920s–1930s, individuals such as Ronald Aylmer Fisher (1890–1962) began to piece the genetics puzzle together. The "modern synthesis" culminated in the late 1930s and 1940s with considerable effort spent understanding implications and applicability of Mendelian patterns of inheritance in the context of Darwinian evolution.

During this time, embryology and the larger field of developmental biology were continuing to develop but with a more experimental and mechanistic bias. In both genetics and development, model systems, such as the fruit fly *Drosophila melanogaster*, were being cultivated. Although, these systems revealed developmental abnormalities due to mutation, the full significance, mechanistic cause, or evolutionary implications of these mutations were not fully understood. One researcher examining such abnormalities, Richard Benedict Goldschmidt (1878–1958), was, arguably, the first to integrate development, evolution, and genetics. Goldschmidt worked on both fruit flies and the nematode worm *Caenorhabditis elegans*, and noted that rather startling developmental abnormalities could take place. For example in flies, a leg could grow where an antenna should be, or a second set of wings develop when there should only be one pair. Goldschmidt coined the term "hopeful monsters," and argued that although rare, there is the possibility that such abnormalities could be favored by natural selection. Goldschmidt encountered serious resistance to his ideas because he was proposing that evolution could proceed in large steps, or macroevolutionary jumps. This notion was contrary to the prevalent view that evolution had to be gradual, as put forth by Darwin himself.

RECONCILIATION OF EVOLUTION AND DEVELOPMENT

Starting in the 1970s, when scientists began to have a better understanding of DNA and how genes work, they began to realize that Goldschmidt's ideas had some merit. We now know that the major developmental abnormalities in Goldschmidt's "hopeful monsters" were actually mutations at a single DNA base pair, a nucleotide, in regulatory regions of a class of protein-coding genes called transcription factors. Transcription factors are proteins that turn groups, or cascades, of genes on and off. If you think of genes as components in an electric circuit, for example, transcription factors are akin to master control switches that turn power to circuits on and off. Modern insight into Goldschmidt's observations, among other developmental discoveries, helped lead a renewed interest in the intersection between evolution and development. However, this time, instead of developmental observations being used to support evolutionary theory (as it was for Darwin, Huxley, and Haeckel), evolutionary theory was applied to developmental observations. The renewed synergy between development and evolution over the past 20 years has resulted in the emergence of a novel sub discipline of biology dealing with the evolution of developmental mechanisms, called "evo-devo" for short.

Several people have contributed to this new area of study, but two in particular deserve special mention. Stephen J. Gould (1941–2002) was a paleontologist by training and is well-known for contributions to modern evolutionary thought and popularization in public venues of evolution. Gould never conducted developmental work and his contributions to the emerging field of evo-devo were theoretical in nature. Gould liked to read and he liked to write. In particular, he took the time to read works of 19th-century scientists (especially the ones mentioned here) and re-examined their ideas in light of modern evolutionary theory and observation. His 1977 book, *Ontogeny and Phylogeny*, was a critical and significant reevaluation of the Haeckelian notion that "ontogeny recapitulates phylogeny."[18] Gould also revisited ideas of heterochrony, or differences in developmental timing, first elucidated by Richard Owen prior to *Origin of Species*. By re-examining the theoretical underpinnings and implications of these ideas, Gould showed that there were several interesting aspects on the evolution of developmental processes that needed further examination.

If Gould provided a theoretical re-synthesis of evolution and development, Rudy Raff was one of the primary researchers promoting integration from an empirical perspective. Raff is a developmental biologist at Indiana University who is also interested in paleontology and evolution. Much of his empirical work has focused on the early development of sea urchins and their relatives. Notably, Raff and his research group have undertaken comparative studies of developmental processes using modern molecular tools. One of the nuances that sets Raff's work apart has been his interest in phylogeny, or evolutionary relationships, of animals. This approach has allowed him to make specific hypotheses about how structures or developmental mechanisms in different species are related. By knowing the relationships between animals, one can more precisely determine which features of embryos and larvae are shared by common descent (homology), which have experienced changes in timing (heterochrony), which have changed their spatial relation to other parts (heterotopy), etc. Once these details are known, one is more able to elucidate mechanisms by which these changes came about (i.e., what were the evolutionary mechanisms that caused developmental changes?).

A major area of intellectual advancement that evo-devo can be credited with is our understanding of how animals are put together. Organisms have an amazingly wide diversity of shapes and forms. Within animals, we have traditionally grouped organisms with a similar form, or body plan, into major taxa called *phyla*. Prior to the advent of recent evo-devo research, most scientists expected that animals in different phyla might have different genes controlling their development and accounting for different body shapes. We now know that this view was incorrect. Within a major lineage, a single group of related animals might show a wide disparity in body plans. Thus, grouping by body plan does not always work.

Second, and more unexpected, even animals with radically different body plans use the same set of genes to control development. Thus, in a fly, a mouse, and an earthworm, the same genes are generally responsible for laying down the general architecture that makes the organism. What varies is when and where those genes are turned on.[19] For example the gene Pax6 is associated with making photoreceptors in a wide variety of organism from flatworms to humans. Pax6 is a conserved transcription factor that if present during embryogenesis will turn on other genes that make a photoreceptor. In some animals, the photoreceptors will develop into very simple organs (as in flatworms), but in others they are very complex (as in humans or squid). Differences in genes that Pax6 turns on and off determine if a complex eye or a simple eyespot forms, but the initiation of Pax6 and the early steps of the gene cascade are very similar across animals. A second example is the gene Distalless, which is found in most if not all animals. In vertebrate animals, it helps specify where limbs form, but in butterflies it helps specify eyespots as well as limbs. Of relevance to human biology, we have discovered that many genes associated with disease or health conditions tend to be evolutionarily conserved and biologically important genes.

The emergence of evo-devo and several other fields of biology have been driven by technical advances in molecular biology and computers in the last 20 years. The use of the polymerase chain reaction (PCR) for specific amplification of genes of interest truly revolutionized biology.[20] As a result, there have been numerous "recombinant" technologies invented that allow detailed observation and experimentation of how genes, sets of genes, and the products of these genes act in different settings. For example, we know several important transcription factors are turned on in very early stages of development. With these tools we can ask, "Does that gene have the same function or effect in 16-cell embryos of related species?" These new techniques also include the ability to sequence whole genomes and expression patterns of all the genes in an organism. Such information is very powerful for understanding how developmental mechanisms have responded to different evolutionary factors. At present, new technologies allow tremendous enhancement of our ability to obtain enormous amounts of genetic data very quickly. Because of the long and rich history between developmental and evolutionary biology, there is no shortage of interesting and significant questions to examine with such approaches. ☙

NOTES

1 Darwin, C. 1859. *Origin of Species by Means of Natural Selection or the Preservation of Favoured Races in the Struggle for Life*. Mineola: Dover Publications, 2006; The Complete Work of Charles Darwin Online. http://darwin-online.org.uk/. Darwin's works are accessible though several web portals, which proved invaluable in preparation of this work. "The Complete Work of Charles Darwin Online" was used heavily from October to November 10, 2009. This site also provided access to Darwin's letters, including the Francis Darwin volumes mentioned in citations 7 and 11.

2 Darwin, 1859.

3 Gladwell M. 2008. *Outliers, The Story of Success*. Boston: Little, Brown and Company.

4 Much of the information on individuals here has been cobbled together over the years during my evolutionary research. However, facts were confirmed through various sources. Wikipedia proved to be very useful and very accurate for much of the historical information on scientists. Those interested in more information on persons mentioned here should see http://www.wikipedia.org and search the person's full name. Most of my use of Wikipedia occurred in late October to November

10, 2009. Birth and death dates were also confirmed with *The American Heritage Dictionary*, 2nd ed.

5 Richards, R. 2008. *Tragic Sense of Life; Ernst Haeckel and the Struggle over Evolutionary Thought*. Chicago: University of Chicago Press. Brief synopses of Richard Owen and Karl Ernst von Baer are given in Richards (2008).

6 Chambers, R. 1844. *Vestiges of the Natural History of Creation*. London: John Churchill.

7 Darwin, F., ed. 1887. *The Life and Letters of Charles Darwin Vol. 1–3*. London: John Murray. See Darwin to Hooker letter.

8 Darwin, 1887.

9 The Alfred Russel Wallace Page. http://web2.wku.edu/~smithch/index1.htm. More information on Wallace can be obtained at "The Alfred Russel Wallace Page."

10 Darwin, 1859.

11 Darwin, 1887.

12 Richards, 2008.

13 Darwin, 1859.

14 Richards, 2008. This biography of Haeckel proves an interesting and entertaining read and provided material herein. It also develops the role Darwin played in Haeckel's life. In particular, some of Richard's interpretations of Darwin's views of recapitulation are non-traditional but worth consideration.

15 Darwin, 1887.

16 Dawkins, R. 2006. *The God Delusion*. Boston: Houghton Mifflin.

17 Haeckel, E. 1874. *Anthropogenie; oder, Entwickelungsgeschichte des Menschen*. Leipzig: Wilhelm Engelmann.

18 Gould, S. 1977. *Ontogeny and Phylogeny*. Boston: Harvard University Press.

19 Carroll, S., J. Grenier and S. Weatherbee. 2004. *From DNA to Diversity: Molecular Genetics and the Evolution of Animal Design*, 2nd ed. Malden: Blackwell Publishing. This book is a great introduction to the field of evo-devo and is filled with examples of some of the important concepts in the field. Both Pax6 and Distalless are covered in this text.

20 PCR, or the polymerase chain reaction, is a laboratory procedure that allows for amplification of very specific regions of DNA. The specificity allows researchers to target a particular gene or gene region for additional study. Furthermore, PCR produces enough of the region of interest to facilitate subsequent laboratory procedures like cloning or sequencing. For a general description, interested readers should visit http://en.wikipedia.org/wiki/Pcr.

Note: This work is AU Marine Biology Program contribution #94.

11

Darwin and the Origin of Life

Anthony Moss

There is irony in including a chapter on the origin-of-life problem in a volume devoted to the contributions of Charles Darwin because Darwin himself despaired of ever understanding the origin of living things: "It is mere rubbish, thinking at present of the origin of life; one might as well think of the origin of matter."[1]

Not that Darwin didn't have insight into processes that led to the first life forms:

> . . . if (and Oh! what a big if!) we could conceive in some warm little pond, with all sorts of ammonia and phosphoric salts, light, heat, electricity, etc., present, that a protein compound was chemically formed ready to undergo still more complex changes, at the present day such matter would be instantly devoured or absorbed, which would not have been the case before living creatures were formed.[2]

Darwin thus revealed himself to be a practical genius: by focusing on the mechanisms of evolution, he was able to make the most of his accumulated experiences. As we will see at the end of this chapter, his carefully researched evolutionary mechanisms are as applicable to chemical evolution as they are to biological evolution.

A Progression of Origin-of-Life Concepts

When creation stories focus on the beginnings of life, they invoke the concept of spontaneous generation (i.e., the formation of life forms from nonliving forms). Ancient Romans believed in spontaneous generation, but were not the only ancient societies to do so.[3] For instance, an ancient Egyptian legend states that the universe and life on Earth originated from nonliving substances in an ocean.[4]

The ancient Greeks also believed in spontaneous generation. The Ionian School speculated that life originated from water and simple, Earth-derived substances. For instance, Xenophane proposed that all life was based on mud.[5] Thales, as described by Pliny, proposed that life originated from slimy substances that were acted upon by light, heat, and air, and that the fundamental element was water.[6] Anaximander proposed that life arose from ancient deep-sea oozes and that all animals and humans arose from fishlike predecessors.[7] Aristotle, in his *History of Animals*, explained that at least the lower animals were derived from slimy muds.[8]

Spontaneous generation, also called abiogenesis or biopoesis, persisted into the early 19th century.[9] During the 1600s, noted fellow and president of the Royal Society Robert Hooke believed in spontaneous generation. Jan Baptiste van Helmont, the 16th-century Dutch scientist famous for his work on willow tree growth that first established the importance of carbon fixation by photosynthesis, was also a follower of spontaneous generation. Van Helmont held that placing sweaty clothes and wheat in a partially closed large openmouthed jar and waiting three weeks could generate mice.[10] In 1668, only 24 years after Van Helmont's death, Francisco Redi provided compelling evidence against the spontaneous generation of maggots on spoiled meat by showing that rotting meat would not grow maggots when covered with a fine muslin cloth that prevented the entry of flies. Redi's work was ignored, and spontaneous generation held sway for another 300 years.[11]

William Harvey (who described blood flow in the early 17th century), Jean-Baptiste Lamark (who proposed an incorrect mechanism for evolution in the early 19th century), and Louis Agassiz (the preeminent American invertebrate biologist in the mid-19th century), all believed in the formation of life by special creation.[12] Even the founders of the cell theory in 1838–39, Matthias Jakob Schleiden and Theodor Schwann, did not accept that all cells necessarily had a precursor parent cell.[13] It was up to Rudolf Virchow to state that all cells arose from previous cells (*omnis cellula e cel-*

lula).[14] August Weismann's germ cell theory stated that complex, sexually reproducing organisms reproduce via a specialized cellular lineage, and he pushed the concept of cellular basis for all cell life back to the beginnings of life.[15] The concept of abiogenesis persisted even though in 1765 Lazzaro Spallanzani had shown that boiling of sealed containers containing microbe-rich water protected against further growth.[16]

In 1859, the French Academy of Science opened a contest designed to settle the issue of abiogenesis. Louis Pasteur entered the competition. His method involved killing microbes in flasks protected against further contamination. To achieve this, he used "swan neck" flasks containing nutrient-rich boiled beef broth. The curved necks of the flasks prevented dust from contaminating the liquid. Broth in undisturbed flasks remained sterile for the length of the study. However, if Pasteur tilted the flask so that the fluid touched the dust and allowed the broth to settle back into the flask, it became cloudy with microbial growth within a day or so. Similarly, if he broke the flask neck, the broth grew microbes; thus, spoilage of preserved foodstuffs arose from the activity of microbes in the environment, not by abiogenesis.[17]

Pasteur's studies gave spontaneous generation its *coup de grace*. But in fewer than 75 years, a new concept of spontaneous generation developed that dealt specifically with the origin of life. (For a more complete historical and scientific examination of abiogenesis, see Iris Fry's excellent *The Emergence of Life on Earth*.[18])

A New Concept of Abiogenesis

The 200-plus years spanning the 17th to the early 20th centuries were heady times for science. Biology was forming the foundation for the modern synthesis—a synthetic view of evolution that developed in the 1930s and 1940s and combined population genetics, developmental biology, and paleontology (see Halanych, Chapter 10). Mendel's systematic studies of pea plant inheritance (1864) revealed the "unit characters" (now called genes) that are the basis of inheritance.[19] Although lost for years and rediscovered just after 1900 by Hugo de Vries, Mendel's work set the stage for the discovery of DNA as the molecule of inheritance.[20]

Archibald Garrod first made the connection between genetics and metabolic disease when he learned that black urine disease followed a Mendelian pattern of inheritance.[21] In the 1920s, Frederick Griffith demonstrated the presence of a "transforming substance" that could be transferred between dead and live bacteria.[22] The "transforming substance" was shown in 1944 to be DNA.[23] Horace Freeland Judson details advancements in molecular biology that occurred from the early 20th century to the mid-1970s.[24] The tremendous technical and conceptual developments of the late 19th and early 20th centuries set the stage for an enhanced understanding of the chemical nature of biology and for the very nature of life.

What Is Life?

All cells are surrounded by a plasma membrane and, thus, exhibit compartmentation, which is a critical requirement for life (the cell itself is a compartment, containing smaller compartments called organelles). Cells undergo self-replication, capture resources from their environment, and have remarkably similar metabolisms—metabolism is the total of all the biochemical activities of a cell—based upon virtually identical biochemical molecules. All living things also have a DNA based information system. All organisms maintain a highly complex homeostasis using exquisitely tuned organic molecular mechanisms (regulatory mechanisms that maintain constant conditions in living organisms). All cells display extremely complex macromolecular components— Proteins, nucleic acids, carbohydrates and lipids—that are indispensable for life. But are all of these things—found as they are today in modern, living cells—critical to prebiotic macromolecular chemistry?

Probably not! Prior to cellular life there were probably organisms that had significantly different chemical and physical characteristics but were capable of self-reproduction and maintenance of simple homeostatic mechanisms. Norman Wingate Perie, as pointed out by John Desmond Bernal, was distressed by the concept of using the term "life" in borderline cases because even clearly nonliving systems have some characteristics of life.[25] Gerald Joyce, working closely with NASA's search for extraterrestrial life, has stated, "Life is a self-sustaining chemical system capable of undergoing Darwinian evolution."[26] As Michael Morange has pointed out, this was stated more completely in 1952 by the British biochemist John Perrett: "Life is a potentially self-perpetuating open system of linked organic reactions, catalyzed stepwise and almost isothermically (i.e., without changing temperature) by complex and specific organic catalysts which are themselves produced by the system." His comments are frequently taken out of context; he does not mean that life can operate independently of its surroundings but that it can use components in the environment to keep the life chemistry going.

Morange warns that scientists and philosophers must be careful not to lock their concepts of life to current technology and understanding.[27] He points to Friedrich Engels (who codified communism with Karl Marx), who referred to life as consisting of entities that could self-replicate, but that those entities were made up of protein. This was most likely because proteins were, in Engel's day, the only biological macromolecules that were somewhat understood. Similarly, Thomas Henry Huxley stated that protoplasm, the "stuff of life," hardly changed from one organism to the next. Although identifying protoplasm as cellular, Huxley had no idea of the true nature of the fluid compartments of the cell.[28] Morange lists three distinct things that all scientists can agree on that are associated with life: 1. reproduction; 2. molecular complexity; and 3. metabolic activity that builds and replicates molecular complexity.[29]

It is difficult to find scientists who will agree on which of the characteristics is most important; most would add compartmentation as well. Recently, Carol Cleland and Christopher Chyba argued that without an understanding of alternative life forms, it is impossible to produce a meaningful definition of life.[30] They make a good point.

Modern scientists need to mimic John Burdon Sanderson Haldane's greater insight. He stated: "All life is characterized by a fundamentally similar set of chemical processes arranged in very different patterns."[31] Our understanding of cellular and molecular function is vastly advanced relative to 1949, but the value of Joyce's viewpoint and Haldane's definition is that it presents a more mature and advanced viewpoint of life as a complex set of functions that are not dependent on specific molecules.

Morange builds his own understanding of the nature of life from Patrice David and Sarah Samadi, who state that living systems have a unique, complex molecular construction maintained by metabolism, which draws upon the environment for its source of molecular components.[32] Life is a series of reactions that display dynamic characteristics that maintain concentrations of reactants, products, electrolytes, and macromolecules (such as proteins, carbohydrates, lipids, and nucleic acids) at nearly constant concentrations between and among the cells of an organism. This steady-state condition (chemically stable but dynamic and responsive) must continue to exist or each individual loses its identity. Under steady-state conditions, the atomic composition constantly changes so that over time the organism becomes constructed of entirely different atoms. The molecules may vary somewhat but they tend to remain at a relatively constant concentration; this is very different from equilibrium. In this steady-state view of life, living systems resist changes at the physiological level by maintaining homeostasis. Yet, we know that living systems incorporate change to a limited degree in their information system—a chemical system intimately connected to the physiological condition via the central dogma, which relates the fundamental flow of molecular and cellular information from DNA through RNA to protein (see Figure 11.1).[33] This incorporation of change in populations over time is evolution.

Morange notes that David and Samadi overlooked evolution in their definition of life, and yet it is a critical part of life. He therefore modified their definition to include the statement that reproduction occurs with variation in the progeny.[34] If this latter characteristic is included, then evolution may occur in any population. Darwin's mechanism of natural selection could therefore serve as a functional description for any—even an alternative—form of life. As we will see below, the capacity for "reproduction with variable products" is necessary even for non-organic, prebiological models of life. In this manner, life takes on a functional, as opposed to structurally dependent, meaning. This brings us back, again, to Haldane and Antonin Oparin.

Chemical Models for the Formation of Life

In the 1920s, Haldane and Oparin independently proposed a strictly chemical-based origin for living systems.[35] Oparin proposed that very simple carbon-based molecules, thought to be in the ancient Earth atmosphere, could undergo reactions to form a relatively dilute "chemical soup" ocean. In some very small-scale environments, locally high concentrations of molecules would generate colloidal micelles (a tiny hollow sphere of lipid;

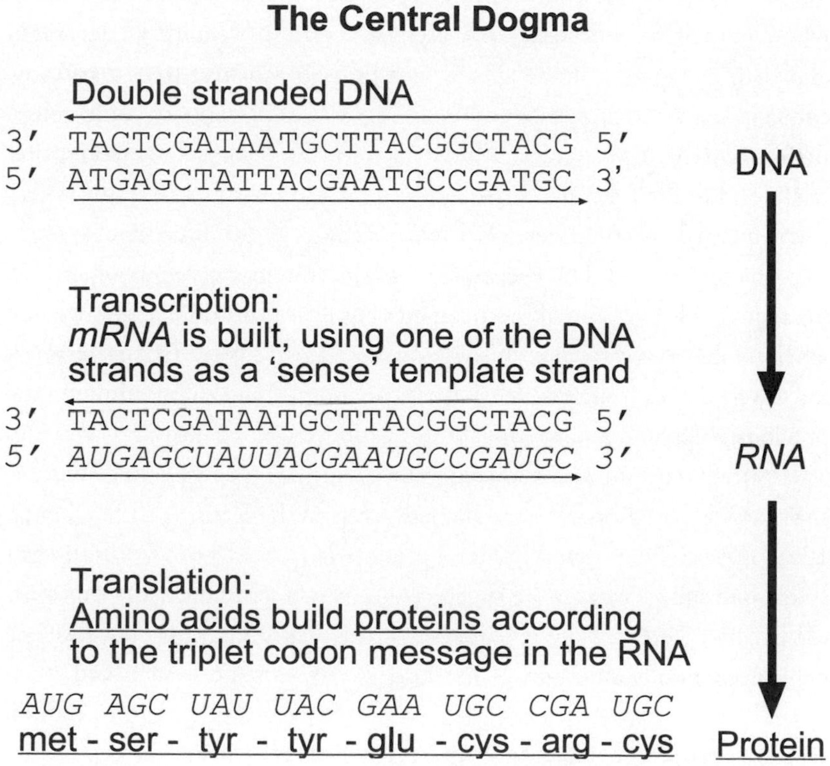

The Central Dogma

Double stranded DNA

3′ TACTCGATAATGCTTACGGCTACG 5′
5′ ATGAGCTATTACGAATGCCGATGC 3′

DNA

Transcription:
mRNA is built, using one of the DNA strands as a 'sense' template strand

3′ TACTCGATAATGCTTACGGCTACG 5′
5′ AUGAGCUAUUACGAAUGCCGAUGC 3′

RNA

Translation:
Amino acids build proteins according to the triplet codon message in the RNA

AUG AGC UAU UAC GAA UGC CGA UGC
met - ser - tyr - tyr - glu - cys - arg - cys

Protein

Figure 11.1

The central dogma: Double-stranded, double-helical DNA harbors the collection of genes of the organism, and acts as a "book." During transcription, messenger RNA is produced. This process transforms the information in the DNA into a form that can be used to produce proteins by the process of translation. Note that the two strands of DNA are of opposite polarity, indicated by the arrowheads, which point to the 3′ ends. Structural information never flows back from protein through RNA to DNA.

a form of colloid). Oparin, in particular, envisioned that colloids—collections of partly hydrophobic, partly hydrophilic molecules surrounded by water but not in solution—would in turn stick together to form complex coacervates that included many different molecules.

Oparin envisioned coacervates to be fluid-like self-reactive masses, which he distinguished from solid, relatively inert, nonspecific aggregates. As a result, in the high-temperature, anaerobic environment of ancient Earth, they could form even more complex molecular assemblies. They would both attract and selectively exclude certain molecules. Coacervates and their component molecules would be very sensitive to pH and ion concentration and would be responsive to heat. Coacervates could select their component molecules, which through intermolecular interactions could build a higher-order, supermolecular structure in some ways similar to cytoplasm. Over vast periods, Oparin's coacervates would give rise first to prebiological and then eventually biological macromolecular chemical machinery. He conceptualized detailed chemical conditions that have not yet been observed. However, his strictly mechanistic approach set the stage for experimental testing of his hypothesis, something that had not been previously possible in any concept of the formation of life.

Entirely without the knowledge of Oparin's ideas, Haldane wrote a short article based on a lecture he had given to the Rationalist Society in 1929. His brief monograph, which predated Oparin's 1934 Russian text, stated in a simpler and more succinct manner much the same scenario as Oparin had. The Haldane-Oparin hypothesis has been the driving concept behind essentially all origin-of-life studies.

Replication First or Metabolism First?

There are two fundamental views of how life may have originated on Earth. There is a "replication first" view, where emphasis is placed on the formation of a replication system. In this model, molecules are sought that are the fundamental basis for molecular and ultimately cellular replication. The RNA world hypothesis, discussed below, fits well into this view. RNAs could be among the first cellular molecules to drive prebiological chemistry while fulfilling the capacity for replication. We will return to a more complete discussion of an RNA-centered view of cellular evolution toward the end of the chapter.

The alternative view is a "metabolism centered" or "metabolism first" model, exemplified by small-molecule chemistries that scavenge environmental energy, replicate, and increase in complexity by coupling energy-releasing reactions with energy-incorporating, cyclic reactions that could form increasingly complex molecular constructs over time. This approach leans heavily on our knowledge that this is in fact how all cells carry out their moment-by-moment functions, in a process that couples molecular breakdown (to release energy) with molecular building (which stores energy). Most of the schemes described here more closely fit this latter "metabolism first" model.

Testing the Chemical Soup Hypothesis

In 1953, graduate student Stanley Miller carried out the first and still most famous of a series of experiments designed to reproduce the early Earth atmosphere and to generate mechanistically a prebiotic chemistry.[36] His experimental design was based specifically upon a proposal made the year before by his geochemist mentor, Harold Urey.[37] Miller boiled and refluxed small molecules routinely found in deep space gases—and which were proposed by Oparin to be critical for the formation of pre-life chemistry: water, methane, ammonia, and hydrogen. Electrical sparks in the atmosphere provided additional energy, including ultraviolet light energy. His experiment was the first to successfully test the Oparin-Haldane Hypothesis, and its results were truly astonishing. In a few days, the Miller-Urey apparatus produced the amino acids glycine, α- and β-alanine, γ-aminobutyrate, and aspartate.

Miller's study stimulated an ongoing series of studies that have generated every major class of complex biological molecule from presumed prebiotic chemical precursors. In subsequent studies, Miller's students and colleagues made nucleotide precursors,[38] nucleotides (the building blocks of the nucleic acids RNA and DNA), which are critical to storing information in modern cells, as well as transforming information, and many other functions,[39] and even carbohydrates—sugars and larger molecules made from sugars, such as starch and the woody stuff of trees and grasses—which are extremely important structural, regulatory, and energetic components of the cell.[40]

Compartmentation

All existing life is cellular; that is to say, all living organisms are based on cells or are single cells. Cells are compartments and themselves contain organelles that are additional compartments inside the cell. Compartmentation is critical because it protects the delicate life-supporting reactions.

Sidney Fox dealt directly with the evolution of compartmentation, although in his case he generated a protein-like membrane, very unlike the lipid bilayer known for biological systems.[41] Fox found that mixtures of amino acids (the building blocks of proteins) heated to 180 degrees Celsius form proteinoids—now called thermal proteins. When warm salty water was added, tiny, hollow, microscopic vesicles—microspheres—quickly formed. Microspheres grew in diameter by incorporating surrounding proteinoids and grew in numbers by budding (the outgrowth and pinching off of small hollow vesicles similar to that seen for yeast cells; a smaller structure grows off the side of a preexisting sphere).

Fox's model was attractive partly because microspheres display decarboxylase and esterase activity.[42] (Decarboxylases remove carboxyl, or -COOH groups, from molecules to typically release carbon dioxide; esterases combine carboxyls and hydroxyls, or -OH groups, of different molecules to link them together with special bonds called esters.) Moreover, they displayed interesting "organellar" dynamics as internal compartments moved about between budding microspheres. In addition, the proteinoid model provided a potential scaffold for the capture of lipids that might eventually form a normal cellular type lipid bilayer. Fox was very successful and trained many students at his Florida State University lab. His ideas were dependent on an easily repeated, robust set of reactions and did not depend on the formation of complex and delicate biological macromolecules. However, late in his career, his increasingly extravagant claims that he had formed early Earth protobionts resulted in sharp criticism and loss of support by the scientific community.[43]

In a recent intriguing series of experiments, 2009 Nobel Prize winner Jack Szostak and coworkers have studied how early precellular structures may have evolved. Szostak's group showed that simple micelle membranes made from fatty acids and glycerol (a small three-carbon molecule, a precursor to fats and many other larger molecules),[44] which are smaller and simpler than phospholipids found in living cells, can be made to encapsulate and retain functional RNA (made up of nucleotide subunit monomers; its structure is nearly identical to DNA) inside artificial protocells.[45] For instance, encapsulated hammerhead ribozymes (which normally cut RNA molecules) retained their activity in Szostak's model protocells.

Szostak's research group has significantly broadened our understanding of what can be done with membrane-bounded artificial protocells. Their protocell model membranes display ion movement, a critical feature of living cells, and some can maintain a gradient of acidity across the membrane.[46] They increase in volume and number[47] and not only incorporate lipids and other materials but can redistribute their contents, including RNA, during division.[48] Sheref Mansy and Szostak have reviewed evidence for how prebiotic cellular types of structures can grow and incorporate critical biological macromolecules. In that model, large spherical vesicles form spontaneously from simple lipids called fatty acids. The fatty acids form multiply layered ('multilamellar') membranes that are less stable than normal membranes; even so, they grow into large, spherical vesicles. The large spherical vesicles spontaneously form long tubular vesicles that, upon gentle agitation, fragment into very small spherical vesicles. These gradually build the large multilamellar vesicles, and the process may continue to cycle in this manner indefinitely. The especially exciting component of this model arises from the observation that such growing and fragmenting vesicles can also incorporate RNA, which in turn can support very simple prebiological chemical processes. An extension of this concept proposes that the fatty acids are gradually replaced with phospholipids. When this happens, RNA and similar large molecules become locked inside the vesicles, finally establishing a stable relationship between RNA and the compartment. Laboratory experiments have shown that these results occur quite easily, and so are very likely to be relevant mechanisms for the eventual generation of true cells.[49]

As successful as this work is, few scientists would expect that RNA or similarly complex and delicate molecules could be the earliest prebiological structures. The ancient Earth would have provided ample quantities of relatively simple rocks and minerals in an extremely challenging temperature environment for such delicate molecules. Accordingly, a number of scientists have thought that ancient prebiotic chemistry, instead, depends upon the reactivity of minerals.

Mineral-Based Pre-Organic Templates and Refugia

In 1949, the physicist J. D. Bernal discussed the potential role of minerals in the formation of pre-organic living systems. More recently, Graham Cairns-Smith proposed that clay could provide the basis for prebiotic replication. Errors in crystal structure could be the basis for mutations. Cairns-Smith holds that replication by crystal growth could be the original basis for prebiotic chemistry. It could even display natural selection (in Cairns-Smith's words, "natural rejection").[50] How or when a transition to organic polymers would be made is unclear. However, Cairns-Smith's model nicely answers Norman Pace's appeal for reason: namely that Earth's temperature(s) during the Hadean Era (from 4.6 billion to 3.8 billion years ago; it is the single prebiotic era of Earth) would entirely preclude cellular biochemistry as we know it.[51] A mineral-based prebiotic model is enticing because of its inherent heat-tolerance.

Clay particles have properties that are compatible with prebiological chemical activity. They are well-known to accumulate small molecules, and this would be a very useful property that would concentrate molecules within the clay structure. Montmorillonite, a common clay mineral, is an excellent catalyst for the formation of RNA from ribonucleotides.[52] Clays also can aid with the formation of membranes.[53] It has further been shown that RNA molecules adsorbed (surface adhesion) onto montmorillonite are protected from UV irradiation.[54]

Crystal faces could be important players in the ancient selection of specific shapes or symmetry of molecules[55] based on variations of optical isomers. All living organisms have L-amino acids while containing D-sugars (although bacterial cell walls contain some D-amino acids).[56] Typically, however, when synthesized in the biochemistry laboratory, amino acids and sugars are produced with nearly equal quantities of both L and D isomers. Robert Hazen's group at the Carnegie Institution for Science and George Mason University showed convincingly that crystalline surfaces selectively adsorb specific amino acid optical isomers. In their experiments, the selective adsorption was weakly biased but repeated. Weakly biased adsorption combined with polymerization resulted in strong overall selection for one specific isomer.

Another mineral-based model has been devised by Gunter Wächterhäuser, a chemist and patent lawyer.[57] His iron-sulfur hypothesis depends on the adsorption of small molecules on iron and sulfur-rich minerals and the release of energy from those minerals as they oxidize to produce minerals related to fool's gold. Iron-sulfur minerals could provide an energy source and a refuge for adsorbed molecules to undergo reactions. Additionally, his model, like Cairns-Smith's, avoids the problem of high-temperature degradation of long organic molecules.

But are minerals like iron compatible with the generation of biologically relevant organic molecules? Recently, Miller's colleagues answered this question by showing that an ancient atmosphere rich in carbon dioxide and nitrogen can support the formation of amino acids in the presence of oxidation inhibitors such as ferrous iron (with a charge of +2; Fe^{2+}, in contrast to ferric iron, which is Fe^{3}).[58]

ENERGY COUPLING

"Metabolism centered" or "metabolism first" models of the formation of prebiotic macromolecules propose that the interactions of small, low-energy molecules over long periods of time cause the formation of steadily larger and more complex molecules that have relatively higher stored energy. This is a fundamental characteristic of life evident in all cells, which must take up materials from their environment to grow and in some cases differentiate. Biological chemistry is marked by a myriad of complex pathways that carry a flow of energy through the organism.

Each pathway is critically dependent on large, delicate molecules, some of which are enzymes (which catalyze and control specific reactions), and also some of which are reactants and others that are products—which are in turn reactants for yet other reactions. Hundreds of different reactions are intimately connected to each other as reaction pairs, with each individual pathway itself critical to proper cell function. In most cases, continued existence of the cell and body is impossible without the ongoing, robust molecular/chemical interactions within each cell. How could such intimate energetic couplings of reactions and the transfer of energy between delicate, unstable molecules occur on a hot, anaerobic Earth, where larger molecules would be expected to thermally degrade?

Christian deDuve recognized that ancient prebiotic chemistries needed a ready source for energetic coupling reactions and that long, extremely complex, high-energy molecules would not hold up under prebiotic Earth conditions. Instead, deDuve, taking advantage of the rapidly growing understanding of sulfur compound chemistry in the 1970s and 1980s, suggested that a common, but typically overlooked, thermostable (stable at high temperature) molecular motif (a "theme" of molecular structure) found within many critical metabolic pathways provides a plausible answer that could contribute to the polymerization of molecules, while itself being involved in important energetic transactions. This molecular group is called the thioester (special unstable bonds that include a sulfur atom), which is formed when sulfur-containing molecules called sulfhydryls (combine a sulfur and hydrogen atom), which would be readily available on an ancient Earth, react with carboxyl groups (combines a carbon with two oxygens and a hydrogen: R-COOH).[59] Thioesters are critical molecules in many biological pathways, including the Krebs cycle and the formation of numerous lipids, amino acids, porphyrins, and many other molecules. Thioester intermediates are critical for the generation of ATP.

The Krebs Cycle is one of the most important metabolic reaction series. It works in a cyclic manner, receiving all of the major classes of biological molecules from the energy-releasing pathways, in order to further squeeze energy out, and feed the next set of reaction series, called the electron transport chain. Porphyrins are complex, square ring molecular motifs, which bind special atoms such as oxygen, magnesium or iron. Porphyrins are commonly colored, and are involved in many critical molecular functions, such as photosynthesis in plants, and oxygen transport such as in hemoglobin in the blood. The "common currency" of cellular energy; it transfers energy via the transfer of phosphates during molecular reactions, and powers most cellular functions.

It is easy to imagine thioesters as molecular "fossils" and predecessors of today's pathways leading to high-energy, large organic molecules.

Another component critical for energetic transformations, molecular signaling, cellular regulation, and all nucleic acids (e.g., DNA and RNA), is phosphate. Inorganic phosphate is the energy transferring group of ATP, made of phosphorus and four oxygen atoms, with 2–3 hydrogen atoms (depending on the location of the phosphate in a molecule). Phosphate became available during the Hadean, a period of heavy extraterrestrial bombardment.[60] ATP, which transfers energy in thousands of reactions each minute within a living cell, contains three phosphate groups. From all that we can tell, the fundamental molecules and groups that support life today through energetic coupling should have been available on a prebiotic Earth.

High Pressure

The diversity of prebiotic molecular polymerization schemes is enormous. In one of his most intriguing experiments, Hazen and colleagues tested whether high pressures such as exists at hydrothermal vents, deep-sea or deep-crust locations could generate prebiotic chemistry. Hazen sealed pyruvate (a small, three-carbon molecule; a breakdown product of glucose that occurs in the cytoplasm) into small inert gold capsules, and found that when crushed under enormous pressure and heated to hydrothermal vent or deep-crust temperatures, the pyruvate polymerized into a complex mixture of long-chain hydrocarbons (molecules formed primarily from carbon and hydrogen; an example would be petroleum).[61] Hazen's work shows that conditions such as those found within the Earth's crust, or at great depths in the ocean, can support formation of organic polymers.

REPLICATION FIRST MODELS: THE RNA WORLD

DNA is very stable and is relatively chemically inactive: it is acted on, but itself rarely acts as a catalyst. Although critical to life, it must be attended by a complex and highly organized, delicate biological macromolecules in order to play its role in the central dogma (see Figure 11.1). Are there other molecules that have information content that are yet active in other ways? In the 1980s an answer to this question was revealed: RNA.

RNA is an interesting and versatile molecule that plays many roles beyond its information-transfer function in the central dogma. There are many different types of RNA in cells; the vast majority are associated with the process of transcription and translation. Studies of RNA in the late 1970s and the 1990s firmly established that a great many RNAs not only have information characteristics but are also structural and that furthermore, many have catalytic properties; these are called ribozymes.[62] Ribozymes form the peptide bond in proteins,[63] regulate telomere length (enzyme that maintains chromosome length),[64] regulate transcription,[65] and do many other cellular functions. An avalanche of results on RNA function in the 1980s and 1990s changed the points of view of many scientists regarding the biological roles of protein and RNA.[66]

Considering how many functions RNA can support, could the earliest life forms have been based primarily on RNA? Many scientists think so. Although first proposed by Carl Woese in 1967, this idea—the RNA world—is usually associated with Walter Gilbert.[67] The RNA world hypothesis is based on the accumulating knowledge of ribozyme function during the 1980s.[68]

As enticing as this concept is, Norman Pace and Robert Shapiro have warned of the danger of thinking that RNAs could be the first identifiable molecules of life. Pace has been concerned because RNA is a notoriously difficult molecule to produce biosynthetically, while Shapiro has argued repeatedly that the formation of large complex molecules as a point for the beginning of life is vanishingly unlikely.[69] Both strongly support the small-molecule, "metabolism first" scenario. The Cairns-Smith and Wächterhäuser models satisfy their concerns well. In those schemes, collections of small carbon-based molecules on the surfaces and interstices of the minerals could result in sufficiently high concentrations to support polymerization and the types of complex reactions discussed above.

After millions of years, basic pre-life catalytic processes that evolved on mica or clay could be turned over to macromolecules like RNA. For instance, if a genetic system could be formed from the twinning of crystals, as suggested by Cairns-Smith[70] those functions could be handed over to an organic molecule system such as RNA, while sequestration into lipid (or some other) vesicle system occurred. RNA and other molecules could start to take over other functions as well, based on their wide range of inherent catalytic capabilities. Complex RNA-based pre-life functions could become more complex and developed over time, but at some point there would again be a period in time when there would be another "passing of the torch" to the next functional molecular generation. The multiple catalytic functionalities of RNA-based ribozymes may have become divided between proteins first produced by the interaction of RNA molecules. Proteins are typically much more efficient organic catalysts than isolated ribozyme (they are the basis for the organic super catalysts called enzymes). In addition, relatively stable DNA would take over as a superior information storage molecule. At some point, RNA would retain its capacity to carry and transmit information as it does today, and living systems evolved to have the tripartite cellular molecular chemistry observed today that underlies the central dogma so clearly demonstrated in all living organisms.

Alternative temperature regimes

Most of the schemes presented so far are dependent on extreme high pressures and/or temperatures. However, Miller and students have carried out long-term experiments, some lasting 25 years, in which they were able to show that the right reagents held at low temperatures can form precursors to complex biological molecules. Sadly, many of the experiments were prematurely terminated shortly before Miller's death because of institutional concern about toxic cyanides (compounds that contain carbon bound directly to nitrogen; they block cellular respiration) included in the reactions.[71]

How could cold conditions drive reactions? It is well-known that reactions slow as temperature drops. Conversely, as concentrations of reactants increase, rates of reaction increase. When ice forms, dissolved salts migrate away from water, forming regions of high solute concentration called eutectic zones. In eutectic zones, solutes are able to undergo reactions because of their high local concentrations. Furthermore, freezing point depression in regions of high salt concentration keeps water liquid even inside masses of ice. It is well documented that molecular synthesis can occur in eutectic zones.[72] Temperature changes may modify reactions. For instance, low temperatures facilitate hairpin ribozyme-mediated splicing (i.e., the attachment of nucleic acid molecules) of RNA—a complete reversal of its usual enzymatic self-cutting activity. Such findings give hope to scientists searching for extraterrestrial life on Saturn's extremely cold moons, Enceledus and Titan. It is unlikely that natural accumulations of solutes are pure, so complex mixtures of molecules, minerals, or clay particles and/or ice could easily combine to form complex precursors (as postulated by Oparin, Wächterhäuser, and Cairns-Smith); as conditions changed over time, a succession of reactions could occur.

The Lost City

In 2000, a new deep-ocean niche for microbial life, based on the formation of serpentine rocks, was discovered by accident at a mid-Atlantic deep-ocean site dubbed the "Lost City."[73] This site is very different from the famous deep-water "black smoker" sites first found on the East Pacific Rise in 1979, where superheated water shoots out of "black smoker" chimneys at >350°C, far above tolerance of any living organism; they support complex ecosystems including invertebrate and vertebrate animal assemblages via symbiotic bacteria that live in the gut of many of the animals. The entire local ecosystem lives off the sulfide-dependent microbial metabolism.[74] Lost City is a moderately deep-water site that shows intense microbial growth based on a warm, but not hot site.

At Lost City the geological formations are very different from those of a black smoker. Lost City forms the mineral serpentine from mantle-derived igneous rock (resulting from cooled lava, called basalt) that is rich in heavy metals such as iron, magnesium, and manganese, called peridotite (a mineral consisting of magnesium and iron bound with silicon, oxygen and water). This fine-grained iron- and magnesium-rich rock is brought close to the seabed surface by the upwelling of magma at the Mid-Atlantic Ridge. Serpentine forms as the peridotite cools and reacts with sea water, a process that releases heat, methane, and hydrogen. Carbonate is also formed as tall white towers, which look like a city. The heated sea water that exits the serpentine rocks is alkaline and very rich in carbonate. When it cools it precipitates fine, lacy carbonate structures that can be up to 60 meters (200 feet) tall. The towers become filled with Archaean microbes that live on methane and hydrogen (see Figure 11.2).

The Lost City site is also different from a typical black smoker because the temperatures are not extremely high—only 40–90 degrees Celsius because the site is approximately 15 kilometers from the hot spreading zone. The Lost City is also different chemically, because the serpentine-forming reactions create a basic pH environment, while black smokers are acidic. Also, unlike in a black smoker, there are few symbiotic associations between microbes and larger animals. Instead, a lacy network of carbonate fills the towers; in turn, the surface is covered with a mixed population of archaea and bacteria, but unlike what is seen at black smokers, few multicelled organisms. Lost City towers are thought to last thousands of years, in contrast to a few hundred years for black smokers.[75]

Lost City has been hailed as another type of environment where ancient prebiotic chemistry could give rise to living cells. This is because in a prebiotic world, such fine, lacy carbonate structures may have formed safe compartments where prebiotic chemistry could safely evolve over time. The long life of these structures could be critical to the formation of ancient cells, also. Of course, we can see only the currently existing organisms. It is unclear how organisms might have originally evolved under Lost City conditions. Understanding this site expands our knowledge of the range of life.

One well-formed overall scheme that seeks to establish a working conceptual framework encompassing many of the features already described, from geochemical models to prelife chemistry to biological chemistry, and which has strong predictive power, is presented by Martin and Russell.

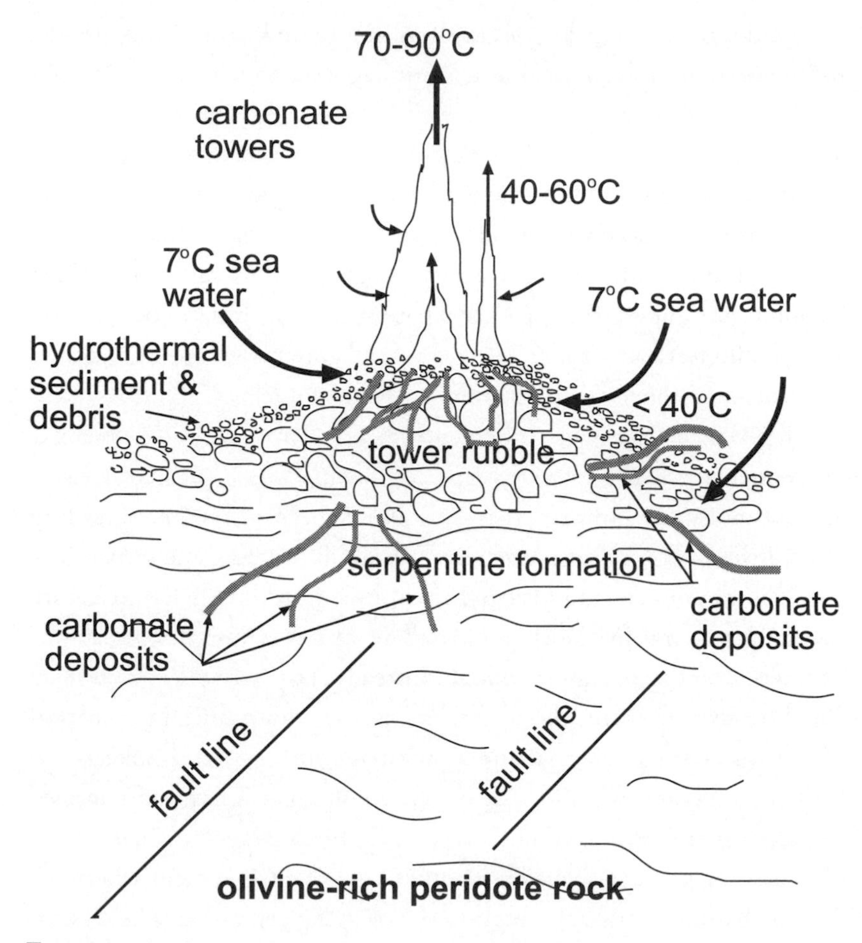

FIGURE 11.2

Organization of the Lost City vent site, found at approximately 3000 meters depth. Peridotite rocks forced upward by Mid-Atlantic vulcanism approximately 15 kilometer from the spreading zone react with seawater to produce serpentine. Water enters the rocks, becomes quite basic and highly enriched in carbonate, which precipitates out when it hits the cold deep ocean water. The precipitate rises above the hot spots near vertically aligned faults in the rock to produce carbonate towers. The serpentine reactions release heat energy, hydrogen and methane, which provide energy to microbes living in the vent towers. The towers and the carbonate deposits are rich with microbe populations.

The reader is encouraged to examine this large, encompassing review for examples of the application of these concepts, and for further insight.[76]

2009: An Exceptional Year for Origin-of-Life Studies

Tracey Lincoln and Gerald Joyce of the Scripps Institute started 2009 with a bang with their report of a self-replicating artificial RNA system that involves two ribozymes that work together to amplify their numbers in solution.[77] The RNA catalysts work together to self-replicate. Joyce and Lincoln built a variant of the 3RC RNA ligase (ligases connect nucleic acid chains), intentionally engineered to make small mistakes in replication. They mixed the different ribozyme variants and added RNA precursor components to build the ribozymes. This scheme makes more ribozymes but with variation built into the products. Thus variation arose not only from the pool of variable reactants but also from mutations that occurred from the inaccuracies built into the replication process. Lincoln and Joyce had purposely set up conditions to try to observe evolution in a population of molecules. Evolution of living organisms is critically dependent upon variation in a population, and is defined as changes in a population in response to natural selection. Variation in a population arises because of mutation. Natural selection acts upon the variation in population, trimming out genetic lineages that are less likely to produce progeny (they are "less fit"). The selection process therefore can act to change the genetics of the population of progeny compared to the parental population; changes in populations with time constitute evolution.

In fact, Lincoln and Joyce's RNA populations did undergo evolution; as the reactions proceeded (thereby producing the next generation), certain RNA ribozyme variants outcompeted others and replicated more effectively, displaying reproductive advantages due to their innate structure. This behavior is exactly what is observed in biological systems. Structural variations led to evolution in the population of molecules, because the structural variation resulted in functional ability to produce progeny. The properties of the populations of RC3 ribozyme variants in the Lincoln-Joyce experiments are analogous to evolution in natural biological populations.

Joyce does not claim to have formed life.[78] He points out that in his experiment the RNAs are, within limits, self-sustaining (i.e., without an external catalyst or other factor) and do produce variation in the progeny. But their system lacks the inherent complexity of biological systems. While the Joyce-Lincoln experiment can produce 144 variants, biological systems are capable of producing many tens of thousands of potential variants, thereby bringing about what Joyce calls "inventive Darwinian evolution." Despite his modesty, it seems clear that Joyce has come close to defining a self-replicating system that mimics life. Perhaps he and his colleagues will soon be able to increase the complexity to the degree required to produce a more lifelike evolutionary model.

One of the most persistently negative features of invoking RNA as a prebiotic molecule is the considerable difficulty researchers have experienced in synthesizing the intact molecule, and its precursors, in the lab. However, in May 2009, John Sutherland and colleagues reported that an unexpectedly simple, temperature-resistant pathway could produce RNA precursors.[79] Their method overcame a fundamental difficulty in the nonbiological formation of the ribocytidine nucleotide (one of four subunit nucleotides of RNA; it is the 'C' of A, T, G and C, and is critical to the formation of RNA). In an unexpectedly efficient and simple scheme, Sutherland's group demonstrated C synthesis under conditions that mimic a prebiotic Earth. In this scheme, phosphate in the reaction acts as a pH buffer, thereby protecting the reactants from degradation. Their experiments were conducted in the presence of UV irradiation at levels meant to simulate early Earth irradiation. The UV destroyed undesired side products and converted some of the ribocytidine to ribouridine, another RNA precursor.

Sutherland's chemistry can therefore account for the pyrimidine precursors, which are one half of the RNA precursors. Michel Nuevo and colleagues have succeeded in producing uracil (a component of uridine) under astrophysical (i.e., outer space) conditions—as might be found on a comet.[80] Uracil thus might be able to be brought to Earth via cometary bombardment. This observation agrees well with the concept that comets were responsible for forming the Earth's oceans.[81] At the time of writing of this chapter, reports of formation of the purine (adenine and guanine) subunits had yet to appear (uridine and cytidine are pyrimidines, with one ring structure; purines have two ring structures).

Is Life *Likely* to Occur?

There seem to be many potential pathways that can produce biological molecules that could lead to the eventual formation of life. Even though there seem to be a wide diversity of conditions permissive to the formation of prelife chemistry, such conditions must persist for a sufficiently long period of time (many thousands to millions of years, which is why the Lost City formations are so attractive). These reactions are very specialized and critics invariably point out that none are being performed under true Hadean conditions (because we don't really know what those were). So, just how likely *is* the formation of life?

Consider an analysis by Eric Chaisson.[82] He points out that the most important characteristic of components within the universe is their associated flow of energy. Not just energy content, but the flux of energy per unit mass: the specific free energy rate, φ_m, which has the units erg/second/gram. Astronomical structures like galaxies and stars have relatively low values, because although a vast amount of energy flows through or is released from them, they are themselves massive. Smaller structures like our planet have a lot of energy passing through a small part of the planet that's relevant to life; this would be the atmosphere and the upper mixed layer of the ocean. In turn, the photosynthetic part of the biological component of the planet captures an enormous amount of energy for its mass, and so has a high free energy rate; in fact it is about 1,000 times that of a star. Higher-order structures like muscle, brain, and other complex parts of animals, and even social components of the biological world, such as civilization components, have much higher free energy rates. A great deal of energy is associated with living systems.

Chaisson asks, "How did this happen?" The answer, fundamentally, is that the universe is not uniform. The universe is not at equilibrium and is constantly expanding. As the universe expands, it cools, and the dust, gas, and other matter of the universe tend to coalesce, forming stars, planets, etc., under the influence of gravity. Since the universe is not only releasing energy but also moving away from equilibrium, there is opportunity for an increase in the local order of the universe (with the coalescence of dust and gases into stars, planets, etc.) at the expense of the energy of the universe. As time passes, matter (and so energy) naturally condenses. Matter seeks out matter via gravitational attraction. Thus, Earth-like planets form. Molecules interact at a much more intimate level and become attracted to mineral surfaces and thereby increase local concentrations of those molecules. Energy that flows through the associated microenvironment (if, let's say, it's a primordial light-gathering mechanism, something that might precede the evolution of photosynthesis), would flow at a very high rate, thereby potentially providing power for complex molecular interactions. The concentrating of a variety of molecules and their resulting interactions may lead toward prebiotic chemistry of the type discussed above. Thus, these types of reactions are simply the natural order and are certain to arise via many evolutionary avenues, potentially giving rise to many different origins of living systems. Chaisson's argument is highly theoretical but hopeful. We may not be able to predict precisely where life may originate. But, if he's right, it's very likely to have occurred multiple times and we are unlikely to be alone in the universe.

Conclusions

In the 200th year of Darwin's birth, it is fitting that we are closing in on plausible origin-of-life scenarios, armed as we are with a stronger understanding of molecular chemistry, guided by Darwin's natural selection. The future is bright for origin-of-life studies, and this author is certain that we will find strongly plausible origin-of-life mechanisms, via continued experimentation, well before the next centenary of biology's most famous son comes around. Multiple plausible origin-of-life scenarios are likely to become evident within the next decade.

Acknowledgments

It is impossible to do justice to all of the researchers who worked so diligently on this inherently difficult topic. My sincere thanks to them for their courage and conviction, and my apologies to those not acknowledged here.

I wish to thank my colleague Jim Bradley and co-editor Jay Lamar for their endless patience while I struggled through this material. I also wish to thank my generous host Martin Zimmer, who allowed me time to write during the summer of 2009. Most importantly, I thank my lifelong love and partner, Ann, for her endurance and patience.

NOTES

1 Darwin, F., ed. 1887. *The Life and Letters of Charles Darwin, Including an Autobiographical Chapter*. The Complete Works of Charles Darwin Online. http://darwin-online.org.uk.

2 Darwin, C. 1871. "*Letter to J.D. Hooker, 1 February 1871.*" Darwin Correspondence Project. http://www.darwinproject.ac.uk/entry-7471

3 McCartney, E. 1920. "Spontaneous Generation and Kindred Notions in Antiquity." *Transactions of the American Philological Association* 51:101–15. http://penelope.uchicago.edu/.

4 Pliny The Elder. ca. 79 AD. *The Natural History, Book IX; The Natural History of Fishes*. ed G.Crane. Perseus Digital Library Project. http://www.perseus.tufts.edu.

5 Oparin, A. 1937. *Origin of Life*. trans. S. Morgulis. New York: Dover Publications, Inc, 1958, 304.

6 Pliny.

7 Kahn, C. 1994. *Anaximander and the Origins of Greek Cosmology*. Indianapolis: Hackett Publishing, 247.

8 Aristotle. 350 BC. *The History of Animals*. trans. D. Wentworth Thompson. London: John Bell, 1907, 304. http://classics.mit.edu//Aristotle/history_anim.html.

9 Brack, A. 1998. *The Molecular Origins of Life: Assembling Pieces of the Puzzle*. London: Cambridge University Press, 428.

10 Oparin, 1937.

11 Redi, F. 1668. *Experiments on the Generation of Insects*. trans. M. Bigelow and R. Bigelow. Chicago: The Open Court Publishing Co., 1909, 33; 36–7. http://books.google.com/books?id=w7ZRAAAAMAAJ&dq=Francesco+Redi+experiment&printsec=frontcover&ct=result#v=onepage&q=linen&f=false.

12 Hawken, P. 2008. *Blessed Unrest: How the Largest Social Movement in History is Restoring Grace, Justice, and Beauty to the World*. New York: Penguin.

13 Stewart, M. 2007. "*Great Ideas of Science: Cell Biology.*" Minneapolis: Lerner Publishing.

14 Virchow, R. 1858. *Die Cellularpathologie in Ihrer Begründung Auf Physiologische und Pathologische Gewebenlehre*. Berlin: Hirschwald.

15 Weisman, F. 1893 *The Germ Plasm. a Theory of Heredity*. London: Scribner.

16 Doetsch, R. 1976. "Lazzaro Spallanzani's Opuscoli of 1776." *Bacteriological Reviews* 40: 270–75.

17 Hazen, R., T. Filley and G. Goodfriend. 2001. "Selective Absorption of L- and D–Amino Acids on Calcite: Implications for Biochemical Homochirality." *Proceedings of the National Academy Sciences* 98: 5487–90; Pasteur, L. 1864. Des Générations Spontanées. In *Louis Pasteur*. Baltimore: Johns Hopkins University Press, 2000.

18 Fry, I. 2000. *The Emergence of Life on Earth. A Historical and Scientific Overview.* New Brunswick: Rutgers University Press.

19 Mendel, G. 1866. *Versuche über Plflanzenhybriden, Verhandlungen des Naturforschenden Vereines in Brünn für das Jahr 1865*. http://www.esp.org/foundations/genetics/classical/gm-65.pdf.

20 Stomps, T. 1954. "On the rediscovery of Mendel's work by Hugo De Vries." *Journal of Heredity* 45: 293–94.

21 Garrod, A. 1923. *Inborn Errors of Metabolism*. Facsimile of the 2nd edition. London: Henry Frowde, Hodder & Stoughton, 293–4. http://www.esp.org/books/garrod/inborn-errors/facsimile/.

22 Griffith, F. 1928."The Significance of Pneumococcal Types." *The Journal of Hygiene* 27:113–59

23 Avery, O., C. Munro and M. McCarty. 1944. "Studies on the Chemical Nature of the Substance Inducing Transformation of Pneumococcal Types: Induction of Transformation by a Desoxyribonucleic Acid Fraction Isolated from Pneumococcus Type III." *Journal of Experimental Medicine* 79: 137–58.

24 Judson, H. 1996. *The Eighth Day of Creation: Makers of the Revolution in Biology*. Cold Spring Harbor: Cold Spring Harbor Laboratory Press.

25 Bernal, J. 1949. "The Physical Basis of Life." *Proceedings of the Physical Society* 62: 597–600.

26 Joyce, G. 1995. The RNA World: Life before DNA and Protein. In *Extraterrestrials: Where Are They?* eds. B. Zuckerman and M. Hart. Cambridge: Cambridge University Press, 139–51.

27 Morange, M. 2008. *Life Explained*. trans. M. Cobb and M. Debevoise. New Haven: Yale University Press.

28 Huxley, T. 1869. "On the Physical Basis of Life." In *The College Courant*. New Haven: Yale University Press.

29 Morange, 2008.

30 Cleland, C. and C. Chyba. 2002. "Defining 'Life." *Origins of Life and Evolution of Biospheres* 32: 387–93.

31 Haldane, J. 1929. "What is Life?" *The Rationalist Annual* 148: 3–10.

32 David, P. and S. Samadi. 2000. *La Théorie De L'évolution: Une Logique Pour La Biologie*. Paris: Flammarion.

33 Crick, F. 1970. "Central Dogma of Molecular Biology." *Nature* 227: 561–3. In Figure 11.1, 3′ and 5′ ends refer to the carbon that binds phosphate at the ends of the DNA strands. The numbers refer to the carbon positions inside the deoxyribose (in DNA) or ribose (in RNA) sugars which are in turn found inside each nucleotide.

34 Morange, 2008.

35 Haldane, 1929; Oparin, 1937.

36 Miller, S. 1953. "A Production of Amino Acids under Possible Primitive Earth Conditions." *Science* 117: 528–59.

37 Urey, H. 1952. "On the Early Chemical History of the Earth and the Origin of Life." *Proceedings of the National Academy of Sciences* 38: 351–63.

38 Oró, J. 1960. "Synthesis of Adenine from Ammonium Cyanide." *Biochemical and Biophysical Research Communications* 2 :407–12; Ponnamperuma, C., R. Lemmon, R. Mariner and C. Melvin. 1963. "Formation of Adenine by Electron Irradiation of Methane, Ammonia, and Water." *Proceedings of the National Academy of Sciences* 49: 737–40; Ponnamperuma, C., R. Mariner and C. Sagan. 1963. "Formation of Adenosine by Ultra–Violet Irradiation of a Solution of Adenine and Ribose." *Nature* 198: 1199–2000.

39 Ponnamperuma, C., C. Sagan and R. Mariner. 1963. "Synthesis of Adenosine Triphosphate under Possible Primitive Earth Conditions." *Nature* 199: 222–6.

40 Weber, A. 2001. "The Sugar Model: Catalytic Flow Reactor Dynamics of Pyruvaldehyde Synthesis from Triose Catalyzed by Poly-l–lysine Contained in a Dialyzer." *Origins of Life and Evolution of Biospheres* 31: 231–40; Zubay, G. 2000. *Origins of Life on the Earth and in the Cosmos.* San Diego: Harcourt-Brace.

41 Fox, S. and K. Harada. 1958. "Thermal Copolymerization of Amino Acids to a Product Resembling Protein." *Science* 128*:* 1214.

42 Nakashima, T. 1987. "Metabolism of Proteinoid Microspheres." *Topics in Current Chemistry* 139: 58–81.

43 Fox, S. 1991. "Synthesis of Life in the Lab? Defining a Protoliving System." *Quarterly Review of Biology* 66: 181–5.

44 Hanczyc, M., S. Fujikawa and J. Szostak. 2003. "Experimental Models of Primitive Cellular Compartments: Encapsulation, Growth, and Division." *Science* 302: 618–22.

45 Chen, I., K. Salehi-Ashtiani and J. Szostak. 2005. "RNA Catalysis in Model Protocell Vesicles." *Journal of the American Chemical Society* 127: 13213–19.

46 Chen, I. and J. Szostak. 2004. "Membrane Growth can Generate a Transmembrane pH Gradient in Fatty Acid Vesicles." *Proceedings of the National Academy of Sciences* 101: 7965–70.

47 Chen, I. and J. Szostak. 2004. "A Kinetic Study of the Growth of Fatty Acid Vesicles." *Biophysical Journal* 87: 988–98.

48 Zhu, T. and J. Szostak. 2009. "Coupled Growth and Division of Model Protocell Membranes." *Journal of the American Chemical Society* 131: 5705–13.

49 Mansy, S. and J. Szostak. 2009. "Reconstructing the Emergence of Cellular Life through the Synthesis of Model Protocells." *Cold Spring Harbor Symposium on Quantitative Biology* 74: 47–54.

50 Bernal, 1949; Cairns–Smith, A. 1988. "The Chemistry of Materials for Artificial Darwinian Systems." *International Reviews in Physical Chemistry* 7: 209–50; 1985. "The First Organisms." *Scientific American* 253: 90–101.

51 Pace, N. 1991. "Origin of Life–Facing up to the Physical Setting." *Cell* 65: 531–3.

52 Ertem, G. and J. Ferris. 1996. "Synthesis of RNA Oligomers on Heterogeneous

Templates." *Nature* 379: 238–40. Ferris, J. 2002. "Montmorillonite Catalysis of 30–50 mer Oligonucleotides: Laboratory Demonstration of Potential Steps in the Origin of the RNA World." *Origins of Life and Evolution of the Biosphere* 32: 311–332; Ferris, J., E. Gözen, V. Agarwal, and L. Hua. 1989. "Mineral Catalysis of the Formation of the Phosphodiester Bond in Aqueous Solution: The Possible Role of Montmorillonite Clays." *Advances in Space Research* 9: 67–75; Joshi, P., M. Aldersley, J. Delano and J. Ferris. 2009. "Mechanism of Montmorillonite Catalysis in the Formation of RNA Oligomers." *Journal of the American Chemical Society* 131: 13369–74; Kawamura, K. and J. Ferris. 1999. "Clay Catalysis of Oligonucleotide Formation: Kinetics of the Reaction of the 5'–Phosphorimidazolides of Nucleotides with the non-basic Heterocycles Uracil and Hypoxanthine." *Origins of Life and Evolution of the Biosphere* 29: 563–91.

53 Hanczyc, M., S. Mansy and J. Szostak. 2007. "Mineral Surface Directed Membrane Assembly." *Origins of Life and Evolution of the Biosphere* 37: 67–82.

54 Biondi, E., S. Branciamore, M. Maurel and E. Gallori. 2007. "Montmorillonite Protection of a UV–Irradiated Hairpin Ribozyme: Evolution of the RNA World in a Mineral Environment." *BMC Evolutionary Biology* 72: 1–7.

55 Hazen, R. 2005. "Ge.Ne.Sis. the Scientific Quest for Life's Origins." Washington D.C.: John Henry Press; Hazen, R., T. Filley and G. Goodfriend. 2001. "Selective Absorption of L- and D–Amino Acids on Calcite: Implications for Biochemical Homochirality." *Proceedings of the National Academy of Sciences* 98: 5487–90.

56 van Heijenoort, J. 2001. "Formation of the Glycan Chains in the Synthesis of Bacterial Peptidoglycan." *Gycobiology* 11: 25R–36R.

57 Wächtershäuser, G. 2007. "On the Chemistry and Evolution of the Pioneer Organism." *Chemistry and Biodiversity* 4: 584–602.

58 Cleaves, H., J. Chalmers, A. Lazcano, S. Miller and J. Bada. 2008. "A Reassessment of Prebiotic Organic Synthesis in Neutral Planetary Atmospheres." *Origins of Life and Evolution of the Biosphere* 38: 105–15.

59 de Duve, C. 1995. "Vital Dust. Life as a Cosmic Imperative." New York: Basic Books; 1995. "The Beginnings of Life on Earth." *American Scientist* 83: 428–37.

60 Pasek, M. and D. Lauretta. 2008. "Extraterrestrial Flux of Potentially Prebiotic C, N, and P to the Early Earth." *Origins of Life and Evolution of the Biosphere* 38: 5–21; Pasek, M. and D. Lauretta. 2005. "Aqueous Corrosion of Phosphide Minerals from Iron Meteorites: a Highly Reactive Source of Prebiotic Phosphorus on the Surface of the Early Earth." *Astrobiology* 5: 515–35.

61 Hazen, 2005.

62 Cech, T., A. Zaug and P. Grabowski. 1981. "In Vitro Splicing of the Ribosomal RNA Precursor of Tetrahymena: Involvement of a Guanosine Nucleotide in the Excision of the Intervening Sequence." *Cell* 27: 487–96; Kruger K., P. Grabowski., A. Zaug, J. Sands, D. Gottschling and T. Cech. 1982. "Self–Splicing RNA: Autoexcision and Autocyclization of the Ribosomal RNA Intervening Sequence of Tetrahymena." *Cell* 31: 137–57; Stark, B., R. Kole, E. Bowman and S. Altman. 1978. "Ribonuclease P: An Enzyme with an Essential RNA Component." *Proceedings of the National Academy Sciences.* 75: 3717–21.

63 Noller, H., F. Hoffarth and L. Zimniak. 1992. "Unusual Resistance of Peptidyl Transferase to Protein Extraction Procedures." *Science* 256: 1416–19.

64 Greider, C. and E. Blackburn. 1989. "A Telomeric Sequence in the RNA of Tetrahymena Telomerase Required for Telomere Repeat Synthesis." *Nature* 337: 331–7.

65 Kilpatrick, M., L. Phylactou, M. Godfrey, C. Wu, G. Wu and P. Tsipouras. 1993. "Delivery of a Hammerhead Ribozyme Specifically Down-Regulates the Production of Fibrillin–1 by Cultured Dermal Fibroblasts." *Human Molecular Genetics* 5: 1939–44.

66 Cech, T. 2002. "Ribozymes, the First 20 Years." *Biochemical Society Transactions* 30: 1162–6.

67 Woese, C. 1967. *The Genetic Code; the Molecular Basis for Genetic Expression*. New York: Harper & Row.

68 Gilbert, W. 1986. "Origin of Life: The RNA World." *Nature* 319: 618.

69 Pace, 1991; Shapiro, R. 2006. "Small Molecule Interactions were Central to the Origin of Life." *Quarterly Review of Biology* 81: 105–25; 2007. "A Simpler Origin for Life." *Scientific American* 296: 46–53.

70 Cairns–Smith, A. and P. Braterman. 1986. "Search for Crystal Genes." *Biomedical and Life Sciences and Earth and Environmental Science* 16: 436–37. Cairns–Smith, A., A. Hall and M. Russell. 1992. "Mineral Theories of the Origin of Life and an Iron Sulfide Example." *Origins of Life and Evolution of the Biosphere* 22: 161–80.

71 Levy M., S. Miller, K. Brinton and J. Bada. 2000. "Pre–Biotic Synthesis of Adenine and Amino Acids under Europa–Like Conditions." *Icarus* 145: 609–13.

72 Kanavarioti, A., P. Monnard and D. Deamer. 2001. "Eutectic Phases in Ice Facilitate Nonenzymatic Nucleic Acid Synthesis." *Astrobiology* 1: 271–81; Trinks, H., W. Schröder and C. Biebricher. 2005. "Ice and the Origin of Life." *Origins of Life and Evolution of Biospheres* 35: 429–45; Monnard, P. and H. Ziock. 2008. "Eutectic Phase in Water-Ice: A Self–Assembled Environment Conducive to Metal–Catalyzed Non–Enzymatic RNA Polymerization." *Chemistry & Biodiversity* 5: 1521–39; Monnard, P. and J. Szostak. 2008. "Metal–Ion Catalyzed Polymerization in the Eutectic Phase in Water–Ice: A Possible Approach to Template–Directed RNA Polymerization." *Journal of Inorganic Biochemistry* 102: 1104–11.

73 Kelley, D., J. Karson, D. Blackman, G. Früh-Green, D. Butterfield, M. Lilley, E. Olson, M. Schrenk, K. Roe, G. Lebon, P. Rivizzigno and the AT3-60 Shipboard Party. 2001. "An Off–Axis Hydrothermal Vent Field Discovered Near the Mid-Atlantic Ridge at 30°N." *Nature* 412: 145–9.

74 Corliss, J., J. Dymond, L. Gordon, J. Edmond, R. von Herzen, R. Ballard, K. Green, D. Williams, A. Bainbridge, K. Crane, T. van Andel. 1979. "SubMarine Thermal Springs on the Galápagos Rift." *Science* 203: 1073–83.

75 Früh-Green, G., D. Kelley, S. Bernasco, J. Karson, K. Ludwig, D. Butterfield, C. Boschi and G. Proskurowski. 2003. "30,000 Years of Hydrothermal Activity at the Lost City Vent Field." *Science* 301: 495–8.

76 Martin, W. and M. Russell. 2003. "On the Origins of Cells: a Hypothesis for the Evolutionary Transitions from Abiotic Geochemistry to Chemoautotrophic Prokaryotes, and from Prokaryotes to Nucleated Cells." *Philosophical Transactions of the Royal Society London* 358: 59–85.

77 Lincoln, T. and G. Joyce. 2009. "Self–Sustained Replication of an RNA Enzyme." *Science* 323: 1229–32.

78 Joyce, G. 2009. "Evolution in an RNA World." *Cold Spring Harbor Symposia on Quantitative Biology* 74: 17–23.

79 Powner, M., B. Gerland and J. Sutherland. 2009. "Synthesis of Activated Pyrimidine Ribonucleotides in Prebiotically Plausible Conditions." *Nature* 459: 239–42.

80 Nuevo, M., S. Milam, S. Sandford, J. Elsila and J. Dworkin. 2009. "Formation of Uracil from the Ultraviolet Photoirradiation of Pyrimidine in Pure H_2O Ices." *Astrobiology* 9: 683–95.

81 Chyba, C. 1987. "The Cometary Contribution to the Oceans of Primitive Earth." *Nature* 330: 632–5.

82 Chaisson, E. 2001. *Cosmic evolution. The rise of complexity in nature*. Cambridge: Harvard University Press.

12

(R)evolution of Man, Pinocchio

From Wooden Puppet to Real Boy

Giovanna Summerfield

Just one year after the death of Charles Darwin, Carlo Collodi wrote his masterpiece, *The Adventures of Pinocchio* (1883), carrying on the revolution against an intervening deity with an emphasis on the competitive struggle for existence and self-betterment, civilized morality, and a new secular order incarnated by Freemasonry. The latter was a fraternal organization with moral and metaphysical values that had grown into a powerful network throughout Europe and the New World since 1717, when the first lodge was established in England. Though both Darwin's and Collodi's membership in this sectarian order is still disputed, the Masons' influence, on personal and intellectual levels, is undeniable. The whole Darwin family was deeply affected by Erasmus Darwin, Charles's Freemason grandfather. Erasmus, an 18th-century physicist, psychologist, and poet, was a high-ranking master of the famous Canongate lodge in Edinburgh, Scotland. He, in a way, laid down the formation for the theory of evolution in which all living things came from a single common ancestor, as explained in two of his books, *Temple of Nature* and *Zoonomia*. Collodi, on his part, was highly influenced by Giuseppe Mazzini, grandmaster of the Grand Orient of Italy (governing body of the Italian Masonry, founded in 1805), and a relentless patriot, otherwise known as a "carbonaro."

Both Darwin and Collodi rejected the explanation of a kind of creation that excludes man's self-management. As the Darwinian theory is based on the law of natural selection and on the adaptability of the human being, so is Collodi's creation and modification of his famous puppet, Pinocchio. Both Darwin and Collodi supported the dynamism of casual and continuous variation, with a tendency toward survival. According to both men, and, for that matter, evolutionists in general, all animals descend from a common progenitor. Present differences are due to changes occurring over time: a bird's wing, a horse's hoof, and a man's arm, though differing in shape and function, were formerly structurally identical. That structure was modified according to differences in nourishment, climate, and setting experienced by each being, as well as by adaptability, which influenced the disappearance of certain characteristics and the acquisition of others.[1]

The essence of Darwin's theory of evolution is the existence of a self-contained system, organized without an external and supreme being or creator, that spontaneously evolves and reproduces. This naturalist philosophy was not entirely new. Similar theories had been advanced by the pre-Socratic philosophers in Ancient Greece, which were embraced by 18th-century figures like Jean-Baptiste Lamarck, whose evolutionism was based on environmental factors. Lamarck's contemporaries and the numerous secret societies of his time, such as the speculative Masonry to which Collodi and Darwin belonged, preserved naturalism.

As James Baldwin attests in *Darwin and the Humanities*, "The theory of natural selection is to be accepted not merely as a law of biology as such, but as a principle of the natural world, which finds appropriate application in all the sciences of life and mind."[2] Baldwin reminds us that *Origin of Species* was followed by *Descent of Man*: Darwin the zoologist was indeed Darwin the humanist.[3] Darwin's theory of evolution, based on radical concepts like over-production with variation, the struggle for existence, survival of the fittest, and the inheritance of characters, challenged these essential concepts—man, nature, God, soul—and placed human condition at the mercy of environment, heredity, and adaptability. Darwin's theory shaped both mental and moral sciences as well as the social sciences and the sciences of language and culture: in psychology, the environmentalist inclination, according to which an individual behavior is the direct consequence of the environment in which s/he operates, grew; in education, humans were seen as evolved creatures that were slowly improving both intellectually and physically as a consequence of interaction;[4] in religion, no divine intervention was considered possible and values and norms were no longer seen as universal truths.

This helplessness and pessimism permeated 19th-century literature and accelerated the decline of the romantic period. In the famous novels of Victorian writers like Thomas Hardy and Naturalist writers like Emile Zola, the main characters appear to be at the mercy of the untamed and untamable forces of nature, largely void of free will or moral choice, shaped only by external factors and the incumbent pressure of circumstances.

Collodi's *Pinocchio* clearly mirrors the Darwinian theory; in Pinocchio's numerous adventures, and eventual metamorphoses, we see a remarkable connection to the key concepts of natural selection as well as its modern adaptations for education, psychology, and literature. *Pinocchio* is a novel of maturation about a piece of pine (an average piece of pine) which becomes, first, a wooden puppet and, finally, a real boy—or better yet, a real "good" boy. Pinocchio's maturation is not only physical but also social; first and foremost, he needs to prove that he is responsible, trustworthy, and hardworking, or, as the blue-haired fairy states in Disney's version, "brave, truthful, and unselfish."[5]

The Adventures of Pinocchio, one of the most famous children's books, is the story of a piece of wood given to Geppetto by Master Cherry. Terribly lonely, Geppetto wants to carve the wood into a puppet to have someone to live with. But this new companion proves very mischievous: both the carpenter and his new "son" land in prison due to Pinocchio's first rebellious act, attempting to leave his adoptive home. When Geppetto is finally freed, he sells some of his garments to purchase books for Pinocchio and sends him to school. But Pinocchio does not seem to appreciate this gesture: he resells everything to attend a theatrical performance. The village puppetmaster, known as Fire-Eater, moved by Pinocchio's remorse, gives him some golden coins to go back to Geppetto. On his way home, Pinocchio falls victim to the trickery of the Fox and the Cat. Saved by the blue-haired fairy, Pinocchio, drawn once more by desire for pleasure, follows a group of young scoundrels to Funland, where, just like everyone else, he is transformed into a donkey. Injured on the job, Pinocchio then is thrown into the bottom of the sea and swallowed by a shark. Inside the belly of the fish, to his surprise, he finds his father Geppetto. Together, they escape while the shark has fallen asleep with his mouth open. Finally back on land, Pinocchio takes care of his father who needs to regain his strength. Pinocchio works while going to school. Then one day, he has a dream: due to his good deeds, the blue-haired fairy forgives and rewards Pinocchio. When he wakes up, he sees Geppetto, steadily working, and, in a corner, the ragged puppet he once was. Pinocchio is now a real boy.

Many scholars have attempted to interpret this famous text from different perspectives. Mark I. West sees Pinocchio's progress as a faithful picture, frame by frame, of a child journeying from adolescence to adulthood, a series of actions and reactions based on the relationship between what Freud calls the "pleasure principle" and the "reality principle." The change that comes after seven chapters of acts of refusal and plain stubbornness culminates in a number of painstaking discoveries about the world and Pinocchio's own dependence. This occurs primarily because he finally follows the dictates of the reality principle. In simpler terms, Pinocchio starts to mature because he is forced to adapt to impending circumstances of which he is aware but cannot control. Pinocchio, like any child, discovers that to attain something one must work for it; he accepts the necessity of working, as any adolescent does, and adjusts to the fact that his own well-being is not due to the efforts of his parents but to the labor of his own brow. Pinocchio becomes productive and enjoys the outcome of his productivity as a young adult does, helping his aged genitor. West insists that the final lesson, then, shows the reward of adulthood against a perpetual childhood that, at first, seems to be gratifying but, eventually, robs him of his own identity by isolating him from those who care for him and by bringing him to a stage of regression and dissatisfaction. Thus, West underscores Collodi's pedagogical agenda.[6]

Jack Zipes, author of *When Dreams Came True*, considers the theme of education complex because "Collodi had not initially planned to allow Pinocchio to develop."[7] It is true that the end intended for the series, originally printed in the children's newspaper *Il giornale per i bambini*, was the puppet's death by hanging. The end we know today was imposed by popular demand on the author. And even this end, according to Zipes, is questionable because Pinocchio is still young and uneducated, although he is willing and ready to take the next step, a step in the dark. Focusing on specifics, like the unique manner in which Collodi opens the book, the selfish reason why Geppetto decides to "conceive" Pinocchio, and the socialization process Pinocchio the puppet undergoes, as well as his physical hardship, Zipes demonstrates that *The Adventures of Pinocchio* is, indeed, a realistic depiction of the condition of poor Italian boys in the 19th century, a representation of the difficulties Collodi himself may have experienced. Zipes is not alone in this speculation. Lois Rostow Kuznets asserts in *When Toys Come Alive* that Collodi's Pinocchio is grounded in the reality of a newly united Italy and that, within this context, the book is an important part of the series of "school books" the author was called to produce for the new nation. According to Kuznets, this purpose dictates the need to endow the puppet with two attributes: a good heart to respond to love, and a wooden body to withstand hardship. The good heart links him to other characters of 19th-century English and American literature, including folklore's so-called Good Bad Boys, who are damned by original sin.[8]

Vigen Guroian, in *Tending the Heart of Virtue*, explores the spiritual character of the Collodian work by stressing that Pinocchio's transformation from wood into flesh is not complete without a good heart: "Collodi is clear that Pinocchio's good heart is the source and substance of his humanity and that responsible relationships with others are that humanity's path to perfection."[9]

Guroian translates a good heart and good deeds into the Catholic tradition with recurrent references to the Bible. He sees Pinocchio's death by hanging as an allusion to the crucifixion of Christ, Pinocchio's meeting with the blue-haired fairy in the Busy Bee Island as the encounter between two disciples and the resurrected Christ on the road to Emmaus, and Pinocchio's encounter with the great shark as Jonah in the belly of the whale. Other scholars before and after Guroian, including the Italian cardinal Giacomo Biffi, have given a Catholic interpretation to Pinocchio's story. But how do we explain that in this little country village in 19th-century Italy neither a religious personage nor a religious event is mentioned, suggesting perhaps that the author deliberately excluded religion? Furthermore, if these critics based their theories on the presence of such values as goodness, generosity, forgiveness, and familial love, one can certainly argue that these are basic concepts of any civil and civic institution.

Guroian adds that in Pinocchio "grace assists but does not compel the moral maturation of the puppet."[10] He is absolutely correct because Pinocchio matures on his own. Others advise and stimulate him, but Pinocchio does not listen. Like a young child who is eager to become independent, he wants to do everything on his own. A celestial intervention is no exception here. Thus, admitting that grace is not included in Pinocchio's plan, Guroian not only weakens his reference to Catholicism but actually dismisses it because in Biblical terms, faith without deeds is dead. The message of the do-it-yourself Pinocchio is not Christian: it is moral, mystical, and a wish of betterment. In fact, it is a kind of writing of the soul that counters deception and represses obedience.[11] Suzanne Stewart-Steinberg claims in *Pinocchio's Effect* that Pinocchio's story marks a moment of crisis in which the individual questions the nature of the social bond and the role of the individual within the society. Within this sense of vacillation, the subjects, a band of brothers at the mercy of but in constant revolt against the patriarchal authority, negotiate within the religious and secular realms. Pinocchio struggles with his right to be independent and to assert his self, sandwiched as he is between

> on the one hand, an older, natural economy founded in the laws of violence (as exemplified by Fire-Eater's theatre), betrayal (the Fox and the Cat), and mendicant artisanship (Geppetto and Master Cherry); and on the other hand, a new economy predicated on the autonomy of behavior founded in wage labor.[12]

The dichotomy extends even further because Pinocchio is not a puppet with strings. He is self-propelled yet still moved by an outside force that he can neither know nor control. Pinocchio proves to be

> a transitional figure, for he represents, on the one hand, the universe of the master puppeteer or the King; and on the other hand, the universe where the King is dead and where power must be exerted in the form of an inner mechanics.[13]

Since there is no king and no patriarchal order in Pinocchio's world and because Pinocchio is both the product and the refuse of a new body politic, one can say that Pinocchio is born into a brotherhood and not into a real family. In fact, he is the incarnation of the Masonic ideology. The proof is in the text itself.

The entire structure of the book is based on three fundamental components of the Masonic creed: liberty, because Pinocchio is a free being who loves his freedom; equality, because Pinocchio's only aspiration is to be like other children and because all characters belong to the same social class; and fraternity, because this is the main feeling that permeates the work. In Chapter 23, we are also introduced to Pinocchio as "fratellino," a neophyte of the lodge on his first degree of initiation. The second degree is represented by Pinocchio being swallowed by the shark, inside of which he finds a candle, a table, and leftovers, a typical ceremonial dressing as described by Cesare Medail in *Corriere della sera*.[14] These plot points are about successfully passing the steps, or initiation rites, while maturing as an individual and as a member of both Masonic society and society at large.

«*C'era una volta un pezzo di legno.*» — "Once upon a time there was a piece of wood,"[15] a piece of wood which happens to appear by pure chance in Master Cherry's cabinet. As an old carpenter, or the Venerable Master of this hypothetical lodge, he sees potential in this piece of vulgar wood. First, he thinks that he could use it as a leg of a chair, but then he realizes that it has an exceptional quality—the wood has a voice and a mind. "At that moment someone knocked on the door"[16] of the temple, and the Master says, "Please come in . . ."[17] Another carpenter, of a lesser degree, comes in. His name is Geppetto.

Geppetto is an odd, irascible old man, who lacks tolerance and patience. The idea of giving Pinocchio to Geppetto is founded on the sense that Geppetto himself needs to be tested in order to change. Geppetto is fundamentally a good friend and a good man, as seen in the first pages of the text and underscored by the author. Furthermore, he is already indoctrinated: he is an apprentice. What is needed is further proof of his commitment to the lodge and further work on his soul, which he can accomplish while sponsoring another apprentice. So Geppetto takes this rough piece of wood into his poor abode, a true "cabinet of reflection,"

> [. . .] an underground, tiny room with a feeble light coming from the stairs. The furniture could not be any simpler: an old chair, a bed in bad conditions, and a table all ruined. On the back wall, one could see a lit-up fireplace but the fire had been painted and close to the fire also painted was a pot, that was boiling gaily, emanating a cloud of make-believe smoke.[18]

where he takes his oath by birthing this new creature.

Thus Pinocchio is born. However, he is not formally finished, and, in spite of this—or due to it—he is already disrespectful and demanding, wanting to take flight whenever he wishes. Geppetto runs after him without being able to catch him; he fears for Pinocchio's susceptibility to being sculpted—completed—by others whose goals might not correspond to the original ones of his puppeteer. Yet, even when physically separated, Geppetto and Pinocchio are connected, almost as if the puppeteer identifies himself with his puppet; they suffer simultaneously and consequently. In Chapter 6, for example, while Geppetto is in prison, Pinocchio faces a cold and violent wind, a bucket of water, and finally the fire that will burn his feet (air, water, and fire).

Geppetto has progressed to the second degree—becoming a Fellow. He has not attained perfection, yet he is not as easily spurred to anger as in the first few chapters. Pinocchio becomes more subservient, more apt than before to display human warmth and a sense of guilt. To convince Geppetto to redo his feet, Pinocchio makes his own oath (it is indeed a test by fire that requires an oath):

> "I promise you"—said the puppet, while sobbing—"that, starting

> from today I will be good . . . I promise that I will go to school, that I will study and make you proud. I promise you, Daddy, that I will learn a skill and that I will be the comfort and support of your old age."[19]

To which Geppetto responds, with devotion and sacrificial love, by selling his coat in exchange for a book for his son.

With rebuilt feet, Pinocchio is faced with another series of tests, still based on air, water, and fire. He risks being burned by Fire-Eater and by the two assassins, the Cat and the Fox; he twirls in the wind while he is hanging and then flies in the sky thanks to a giant dove; finally, he is thrown into the sea as a donkey to be killed and used as skin cover for a drum and then goes in the water again to reach babbo Geppetto, a purifying stage through which he will come out whole and righteous, ready for his total commitment to the lodge, the blue-haired fairy, indeed.

Not always visible but nevertheless present and involved in his transformation, the fairy represents Freemasonry, the ideology that combines charity and rationalism rather than goodness and faith. The number 3 appears constantly in references to the fairy: for instance, when she intervenes for the first time to save Pinocchio's life, while he is hanging from the oak, she knocks three times. Later, while he is recovering in her house, Pinocchio is visited by three doctors. The fairy first appears in the shape of a child, who vows to Pinocchio that he will be her little brother. The second time Pinocchio meets the fairy, she is not a child any longer, and during this meeting Pinocchio declares his desire to become a real boy and understands that he will have to overcome a series of tests to do so. Since his behavior is improving, the fairy announces that the time has come and prepares a party for his metamorphosis, but Pinocchio turns down this chance in order to enjoy the profane world of Funland. After his terrible experience there, Pinocchio's redemption begins. Pinocchio will see the fairy, for the third time, in the shape of a goat, which, in turn, aids him while he is being swallowed by the shark, an event that marks the start of his definite change.

Collodi has followed his protagonist through a series of adaptations: from vegetable to animal, then, within the same natural kingdom, from one animal into another, and, finally, from object into human being. His name alone, which in Tuscany at the time meant "pine nut" or "pine seed"—the contemporary term is «*pignola*»— associates him with a plant. He begins life as a featureless stick of firewood (pine wood), which is then transformed into a wooden puppet. Next he moves through animal identities (watchdog and donkey), and finally achieves full human status. Yet we still see Pinocchio as a puppet sitting floppily there while a little boy admires himself in the mirror, concurring with Darwin's philosophy that "man with all his noble qualities, with sympathy and benevolence, with his god-like intellect, with all these exalted powers, man still bears in his bodily frame the indelible stamp of his lowly origin" and that "in the struggle for survival, the fittest win out at the expense of their rivals because they succeed in adapting themselves best to their environment."[20] The «*ragazzo perbene*,» the upstanding boy, has proven to be the fittest; he is the Masonic brother who has left behind all his weaknesses and who, thanks to his abilities, has won the battles of life, becoming a role model for others. Collodi's Pinocchio was set to demonstrate that human progress, perseverance, and independence, rather than providential design, are the elements needed for a successful existence. ☙

NOTES

1 *I mondi dell'uomo*. 1975. Verona: Arnoldo Mondadori, 176–7.

2 Baldwin, J. 1909. *Darwin and the Humanities*. New York: AMS Press, 1980.

3 Ibid.

4 Morton, W. 1943. *The Origins of Dewey's Instrumentalism*. New York: Columbia University Press.

5 Guroian, V. 1998. *Tending the Heart of Virtue*. Oxford: Oxford University Press, 42.

6 Rollin, L. and M. West. 1999. *Psychoanalytic Responses to Children's Literature*. Jefferson: McFarland & Company, Inc, 65–70.

7 Zipes, J. 2007. *When Dreams Came True*. New York: Routledge, 180.

8 Kuznets, L. 1994. *When Toys Come Alive: Narratives of Animation, Metamorphosis, and Development*. New Haven: Yale University Press, 67–70.

7 Guroian, 1998.

10 Ibid.

11 Stewart-Steinberg, S. 2007. *The Pinocchio Effect: On Making Italians (1860–1920)*. Chicago: University of Chicago Press.

12 Ibid.

13 Ibid.

14 Medail, C. 1993. "Pinocchio? Un Fratellino della loggia di Firenze." *Corriere della Sera*, April 16.

15 Collodi, C. 2002. *Le Avventure di Pinocchio*. trans. C. Collodi. Milano: Garzanti.

16 Ibid. Author's translation of: «*In quel punto fu bussato alla porta.*»

17 Ibid. Author's translation of: «*Passate pure . . .*»

18 Ibid. Author's translation of: «*[...] una stanzina terrena che pigliava luce da un sottoscala. La mobilia non poteva essere piu' semplice: una seggiola cattiva, un letto poco buono, e un tavolino tutto rovinato. Nella parete di fondo si vedeva un caminetto col fuoco acceso ma il fuoco era dipinto, e accanto al fuoco c'era dipinta una pentola che bolliva allegramente e mandava fuori una nuvola di fumo che pareva fumo davvero.*»

19 Ibid. Author's translation of: «*Vi prometto, — disse il burattino singhiozzando, — che da oggi in poi saro' buono [...] vi prometto che anderò a scuola, studierò e mi farò onore [. . .] Vi prometto, babbo, che imparerò un'arte e che sarò la consolazione e il bastone della vostra vecchiaia.*»

20 Darwin, C. 1874. *The Descent of Man*. 2nd ed. New York: A. L. Burt, 619.

Author's Note: Some of the Pinocchio's paragraphs presented here also appear in Chapter 4 of *New Perspectives on the European Bildungsroman* by the same author. London: Continuum, 2010.

13

Darwin's Great Idea and Why It Matters

James T. Bradley

This essay is written mainly for four groups of people:

1. those who do not know very much about biological evolution but are interested in learning about it;
2. those who have some knowledge about evolution but do not believe it is a good explanation for the origin of the diverse life forms on Earth;
3. those who know about evolution and would like to be able to talk about it more easily with others who either know little about it or are adamantly opposed to it; and
4. those who are presently opposed to the theory of evolution.

In other words, this essay is written for all high school and university students, biology teachers, and naturally curious and communicative persons.

Evolution Literacy 150 Years After *Origin of Species*

Public knowledge about Darwin's ideas and modern evolutionary biology is poor. A Gallup poll released in 2009 on the eve of Darwin's 200th birthday reported that only 39 percent of Americans "believe in the theory of evolution."[1] Reasons for rejection of Darwinian theory are multifaceted, but two major factors in the United States are: ineffective teaching in public grade schools of the basic principles of biological evolution, and effective teaching in the churches of several Christian denominations of misinformation about evolutionary theory and its implications. These two factors exist in a synergistic relationship fueled by fear.

On the one hand, public school teachers either omit or give cursory classroom attention to evolution for fear of offending or antagonizing students, parents, or school boards who oppose learning or teaching about evolution. On the other hand, clergy and other religious teachers, primarily in conservative Protestant denominations, often use fear of eternal damnation or misinformed statements to discourage their congregants from learning about evolution from those most qualified to teach about it: biologists.

Fear is also at the root of the anti-evolution positions of some churches—fear that accepting biological evolution as a science-based fact will undermine religious faith and the authority of scripture. R. Albert Mohler Jr., president of Southern Baptist Theological Seminary, claimed in a 2005 interview with *Time* magazine that evangelical Christianity and evolution are incompatible beliefs:

> You cannot coherently affirm the Christian-truth claim and the dominant model of evolutionary theory at the same time . . . Personally, I am a young-Earth creationist. I believe the Bible is adequately clear about how God created the world, and that its most natural reading points to a six-day creation that included not just the animal and plant species but the Earth itself.[2]

So there is a vicious cycle. Fears of church leaders beget congregations which beget fears in public school systems which beget students who grow into adults unequipped to challenge the bases of church leaders' fears. The result is that generations of Americans have a trained incapacity to draw upon principles of evolutionary theory in their professional, civic, and personal lives.

Public, private, and homeschool educators in all disciplines need to do a better job of learning and teaching about biological evolution. High school and university students also bear some responsibility to seek out information about biological evolution beyond what they may have been told in their homes and churches. Students and other citizens of the global village deserve to know what evolution is, what evolution is not, and why knowledge about evolution matters.

What Evolution Is

Evolution is, simply, change over time. We can speak of the evolution of art forms, car and hair styles, and musical instruments like the violin or piano. Biological evolution refers to changes in life forms over time, and these changes are manifestations of changing frequencies of genes in populations over time. That life forms have changed over time—evolved—is a fact. The fossil record is replete with extinct species of plants and animals.

Whether biological evolution occurred is not at issue. The question is: how did evolution occur? Darwin's answer, natural selection, may not be the complete answer, but it is certainly a very important part of the answer, perhaps the major part. Evolution by natural selection occurs due to three facts of life:

(1) Living things produce more offspring than the environment can support. In other words, organisms' reproductive potentials exceed the carrying capacity of the environment.

(2) Inherited variations exist among the individuals of breeding populations. For example, in a population of robins, some birds may be better at hearing worms in the soil, have a brighter orange color to their breasts, build nests faster and sturdier, or sing louder than others.

(3) On average, individuals with physical, physiological, or behavioral traits that give them a reproductive advantage over others leave more offspring in the next generation. This phenomenon is natural selection. Since offspring carry their parents' genes, the frequency of genes for adaptive traits increases in a population. If an environmental change makes different traits more adaptive than those predominating before the environmental change, the genes for those different traits will be selected for, and their frequency will increase in the population. Darwin did not know about genes, but he did know that offspring somehow inherit traits from their parents.

Darwin's great insight was that over immense stretches of time, natural selection can give rise to new species of plants and animals. Emergence of new species from preexisting species is called speciation. Several natural processes can lead to speciation. All involve the reproductive isolation of a subpopulation of breeding individuals from the main population. One way reproductive isolation can occur is by geographical isolation of a subpopulation from its parent population. As depicted in Figure 13.1, this may happen when geological forces create a mountain range, splitting an existing population into two subpopulations, one on each side of the mountain range. Different climates and biological communities exist on opposite sides of a mountain range. For example, the Cascade Mountains in the Pacific Northwest create a rain shadow that results in a desert on the eastern side and lush woodlands on the western side. Exposed to different environments on either side of the mountain range, subpopulations of

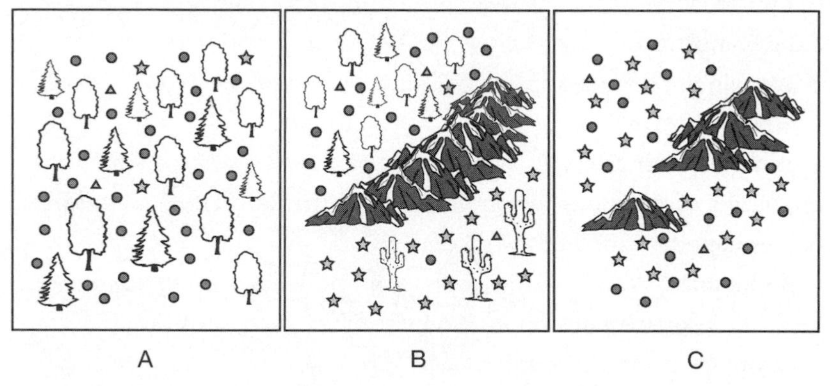

FIGURE 13.1

Depiction of speciation due to geographical isolation. Individual variants in populations are represented by circles, triangles, and asterisks. (A) A population of individuals belonging to the same species lives in a forest where circles are better adapted than triangles and asterisks, so they contribute more offspring to succeeding generations. (B) A mountain range rises and separates the original population into two geographically isolated populations. On the west side of the range, a forest environment remains and circles still predominate; on the east side of the range, a rain shadow causes a desert environment where asterisks are better adapted and come to predominate. (C) Much later, glaciers erode a pass through the mountain range, allowing organisms on each side of the range to mingle; but by now natural selection has produced enough other changes (physiological, behavioral, physical) that the two populations can no longer interbreed. Speciation has occurred.

plants and animals experience different pressures from natural selection and accumulate different adaptive traits. Now imagine glaciations that wear a pass through the mountains, allowing individuals in the previously isolated subpopulations to intermingle. If one or more of their differing adaptive traits makes it impossible for individuals from the two subpopulations to interbreed, speciation will have occurred.

Sometimes subpopulations of individuals from one species become isolated on different islands and give rise to different species. This is what Darwin correctly surmised to have happened with the finches and tortoises on the Galapagos Islands. Modern-day biologists have also documented rapid speciation among fruit flies on the Hawaiian islands and fish in large freshwater lakes such as Lake Victoria in east Africa and Lake Apoyo in Nicaragua.[3]

Darwin recognized that traits distinguishing individuals from each other are inherited, but he had no idea how inheritance happened. Modern genetics and molecular biology have solved that mystery. We now understand the sources of genetic variation upon which natural selection acts. These sources include mutations in the bases of DNA and a shuffling of gene segments that creates new genes with new functions overnight. However, the greatest source of genetic variation is the random recombination and mixing of genes that happens during sexual reproduction. Although gene variants arise largely by chance in a population, we will see below that the paths taken by evolution are not random.

Evolution Is Not "Just" a Theory

Saying that evolution is "just" a theory suggests that evolution is a guess or a hypothesis about how Earth's diverse array of living things came to be. This is a misuse of what scientists mean by the word *theory*.

Scientific theories are about the closest that science can come to factual explanations for natural phenomena. Science progresses by examining the validity of hypotheses via empirical tests and/or observation of nature. The more predictions from a hypothesis that are borne out by experimentation or observation, the surer one can be that the hypothesis is correct. When many different predictions based on a particular hypothesis are tested and retested over a long period of time and shown to be accurate, a hypothesis may be elevated to the level of scientific theory. This process of testing and retesting has given rise to several well-known theories:

- the heliocentric theory for the solar system—the Earth and other planets orbit the sun;
- the germ theory for disease—infectious diseases are caused by small vectors like bacteria and viruses; and
- the atomic theory for matter—matter is composed of atoms, e.g. H_2O being comprised of hydrogen and oxygen atoms.

The theory of evolution (that all life forms on Earth arose from ancestral life forms) is just as certain as the heliocentric, germ, and atomic theories. This is not to say that scientists know completely and exactly how evolution occurs. For example, biologists continuously argue over which sources of genetic variation are most important for evolution and over the relative importance of natural selection, sexual selection (see Chapter 5), and genetic drift in evolution. But the reality of evolution itself having occurred is not controversial among biologists.

Macroevolution Does Not Contradict Darwin

Overwhelming evidence, now bolstered by DNA sequencing data, supports evolutionary biology's claim that all living things share a common ancestry. Yet the outward appearance of organisms, especially to persons not trained in biology, can make this claim difficult to fathom. For example, evolution's evidence shows that fish gave rise to amphibians, which gave rise to reptiles, which gave rise to birds and mammals. Our feathered, flying friends using the birdbath look so different from the turtle leisurely crossing the road or the squirrel perched on the bird feeder that it is hard to imagine all three sharing a common ancestor. Yet they do. On the one hand, the changes that occurred above the species level for an ancestral group of organisms like reptiles to give rise to a very different-appearing group such as mammals is sometimes called macroevolution. Microevolution, on the other hand, refers to very small-scale genetic changes that can produce different gene frequencies within subgroups of a population (e.g., for human stature or hair color) or, over time, groups of closely related species (e.g., the several species of fruit flies on the Hawaiian Islands).

Biologist colleagues that I know personally do not use the term macroevolution, although most evolution textbooks have a section on it.[4] There is no doubt among evolutionary biologists that Darwin's theory of evolution by natural selection accounts for macroevolution. But some controversy does exist over the rate of macroevolutionary changes and the details of the process. Most biologists view macroevolution as microevolution spread out over immense time spans. When relatively rapid and dramatic evolutionary changes are seen in the fossil record, they may be explained by changes in the patterns of activity of genes that control embryonic growth and development. For example, changes in the onset and duration of cell divisions in certain regions of an embryo can result in gross alterations in the morphology of the adult form. Also, changes in just one or a very few genes can alter the placement, size, and/or identity of body appendages. Not all biologists agree on the relative importance of small, stepwise microevolutionary events and more rapid evolutionary changes due to genetic variations that act upon early, embryonic development. But biologists overwhelmingly agree that evolution via natural selection and a few other natural processes sufficiently explain the wondrous diversity of plant and animal life in Earth's biosphere.

Some people who do not accept biological evolution as the source of life's many forms believe that the dramatic, visible differences between different groups of plants and animals (e.g., moss versus redwood trees, fish versus mammals, etc.) cannot be explained by the theory of evolution. They may accept microevolution as the process whereby the various and sundry breeds of dogs arose from a wolf-like ancestor, but they deny that macroevolution could have occurred.

If macroevolution is real, they ask, why have no new phyla or classes of animals recently appeared on Earth? It is true: no new phyla or classes of plants or animals have appeared on Earth since long before humans were around. But evolutionary theory does not predict that this should happen. In fact, it predicts just the opposite.

The tree of life has many large boughs (phyla), many sizeable branches (classes), and millions of tiny twigs (species). As one evolutionary biologist points out, asking why no new phyla have recently appeared on Earth is like asking why no new boughs have recently sprouted from the twigs in the upper reaches of an oak tree in your backyard.[5] Various phyla of plants and animals diverged from ancestral life forms hundreds of millions of years ago. Since then, branches with countless twigs grew from those ancient phylogenetic boughs. Many twigs and branches, and even a few boughs, stopped growing and became extinct. But the descendants of some boughs persist into the present as members of today's rich and diverse microbial, botanical, and zoological worlds.

So evolutionary theory cannot predict macroevolution, in the sense of major new groups of plants or animals suddenly appearing. Instead, it predicts that the fossil record should document the sequential appearance of major groups of living things, generally from simplest to more complex. And this is exactly what is seen: first, single-celled life, then multi-celled invertebrates, and, finally, vertebrates; likewise, within the vertebrates, first came fish, then amphibians, dinosaurs, mammals, and birds.

Transitional Fossil Forms Are Not Lacking

Evolution by gradual changes over time predicts that transitional forms of life should appear in the fossil record. For example, if birds emerged from dinosaurs, there should once have been dinosaur-like animals with wings, and if fish gave rise to four-legged land animals, there should have been fish-like creatures with leg-like appendages. Remarkably, many detractors of evolutionary theory claim that transitional forms are absent from the fossil record, but this is simply not true. The fossil record contains many transitional forms, and modern DNA analyses are revealing more and more about the evolutionary relationships between groups of living organisms.

Whale evolution is a fine example of how fossils and DNA analyses tell us about the transitional forms that led to a modern-day group of animals. Whales are mammals whose ancestors lived on land. Many fossils from the first 10 million years of whale evolution are from aquatic creatures that still had limbs like those of terrestrial mammals. DNA analyses show that whales are closely related to even-toed ungulates (cows, sheep, pigs, and hippos), and a raccoon-sized, terrestrial, fossil ungulate with several whale-like characteristics was described in 2007.[6] A second example of transitional forms is in the evolution of birds from feathered dinosaurs living about 160 million years ago. The famous winged dinosaur-like bird, *Archaeopteryx*, was discovered in Bavaria in 1861. In 2009, scientists unearthed a chicken-sized fossil of a four-winged, feathered dinosaur in northern China which apparently used its four feathered appendages for gliding.[7] As a third example, consider the movement of fish-like creatures onto land. Paleontologists have discovered many transitional forms between fish and four-legged land creatures. One dramatic 2004 find was in northern Canada. *Tiktaalik*, a fossil predatory "fish" with leg-like fins, a flexible neck, and a skull resembling that of a salamander, lived 375 million years ago. It was three feet long and had weight-supporting forelimbs that could bend at the elbow.[8]

Biological Evolution Is Not a Random Process

Genetic variants arise randomly in populations, but natural selection is not random. Pressures exerted by natural selection are directional. If hard-shelled nuts become the main food source for a population of finches, natural selection will produce a preponderance of birds with strong, sturdy beaks good at cracking the nut shells, not dainty, slender beaks better at sipping nectar from flowers. Natural selection is simply the preferential reproduction of individuals with traits best suited for keeping them healthy and alive long enough to successfully reproduce. The name of the game is reproduction, and reproductive fitness is not necessarily the same as physical strength. A big, strong, male duck with aberrant courtship behavior will probably not pass its genes into the next generation of ducks, and the same goes for humans. Evolution moves in the direction of increased reproductive fitness.

No Contradiction of the 2nd Law of Thermodynamics

Some anti-evolutionists invoke the 2nd law of thermodynamics to bolster their belief that Earth's diverse species do not share a common ancestry. They say that evolutionary theory contradicts the 2nd law of thermodynamics and therefore cannot be true. Their claim is based on misunderstanding or misrepresentation of both thermodynamics and evolution. Their mistaken notion about thermodynamics is that highly ordered entities cannot arise from less-ordered systems. And their mistaken notion about evolution is that random genetic variation cannot give rise to increasingly ordered and complex organisms over time without violating laws of nature.

Let's examine both notions.

Thermodynamics is the study of heat and other forms of energy. Specifically, it deals with transformations between different kinds of energy, as when chemical energy in gasoline is transformed into the kinetic energy of a moving car. Two laws about energy and the universe discovered in the 19th century are relevant to this discussion. The 1st law of thermodynamics states that energy can neither be created nor destroyed; in other words, the total energy content of the universe remains constant. The 2nd law of thermodynamics states that every energy transformation occurs so as to result in an increased randomness or disarray *in the universe* (or in a closed system). I've italicized *in the universe* because this is what is usually forgotten about in arguments that claim contradiction between evolution and the 2nd law.

Energy comes in many different forms including kinetic, electrical, nuclear, chemical, matter itself, potential, and heat. The first law means that when one form of energy, such as the kinetic energy in falling water, is transformed into other kinds of energy, such as electrical energy and heat via a turbine at the base of a waterfall, there is no energy lost in the process. However, since heat energy is unusable for performing work, there is a loss of energy useful for driving processes that create order ranging from cell growth to the building of a cathedral. In this example, the amount of electrical energy produced by the turbine will always be less than the kinetic energy of the falling water since some of the water's kinetic energy is transformed into heat energy. This loss of usable energy is happening all over the universe all of the time—as stars burn, as galaxies rotate, and as plants and animals reproduce, grow, and die. The constant loss of usable energy translates into an increased disarray of the universe as a whole. Another word for disarray is entropy. So another way of stating the second law is to say that the entropy of the universe is constantly increasing. The only way this could be avoided is if new energy were constantly being added to the universe. But due to the 1st law, this does not happen.

Now, what is crucial to note is that just because the universe is constantly becoming more disorganized, it does not mean that increasing organization cannot occur at various sites throughout the universe. In fact, we see this happening all around us. Water molecules in snowflakes are more organized than in the vapor from which snowflakes crystallize. Michelangelo carved his exquisite *David* from a crude block of marble, single fertilized egg cells develop into organisms with complex tissues and organ systems, Mustangs and Ferraris are constructed from tiny particles of iron from inside rocks of iron ore, and beautiful buildings are assembled from bricks made from clay in the soil. You get the idea. And evolution is no exception. In Darwin's words, ". . . from so simple a beginning endless forms most wonderful and beautiful have been, and are being evolved."

How can this happen? How can order arise from disorder and not violate the 2nd law of thermodynamics? The answer is that these examples—snowflakes, works of art, developing embryos, cars, buildings under construction, and evolution—are not closed systems. They are open systems, systems that receive an input of energy from outside them that more than compensates for the usable energy lost as heat.

Consider Michelangelo and his block of marble. For months he hammered and chiseled away at the marble, gradually revealing the *David* within. This hammering and chiseling required lots of energy. Chemical energy in Michelangelo's arm muscles was converted into kinetic energy in a swinging hammer. The chemical energy in his muscle cells came from the breakdown of highly organized food molecules like starch in bread and pasta and animal proteins in sausage and fish. Energy stored in the molecular order of starch and proteins ultimately came from radiant energy of the sun used by green plants for photosynthesis. Recall that photosynthesis uses light energy to assemble simple water and carbon dioxide molecules into highly organized molecules of complex carbohydrates. At each step between photosynthesis and Michelangelo's swinging hammer, some usable energy was lost as heat. But thanks to the sun and to the fact that Michelangelo continued eating, the *David* was completed. All of this happened at the expense of order in the universe.

If you add up all of the disorder created by Michelangelo digesting highly organized food molecules and compare that with the order created from the block of marble, the disorder would be greater. Not only the *David*, but life itself, exists at the expense of order in the universe. For the growth, building, and evolution of complex things here on Earth, the necessary input of energy ultimately comes from the sun.

Now what about evolution? Persons who believe that random genetic variation cannot lead to increased complexity fail to recognize the ratchet-like action of natural selection. Once a random genetic event gives an organism a reproductive advantage, natural selection preserves and amplifies that genetic variant. Succeeding variations add to preexisting ones, with natural selection ratcheting up the fitness and complexity of individuals in a population all the while.

Combining this view of natural selection with the 2nd law of thermodynamics, we see how evolution operates within the bounds of physical law. Natural selection constantly eliminates organisms less genetically adapted for successful reproduction. The eliminated die without reproducing, and their highly organized cells ultimately decompose back into water, carbon dioxide, nitrogen, and a few elemental minerals. Of course, the same fate awaits the winners of life's competition, but not until their genes move into the next generation.

Without the sun to provide the energy needed for the growth and development of the many variants upon which natural selection acts, evolution would come to a screeching halt. But the sun does exist! And natural selection does its winnowing, without violating any laws of thermodynamics. All the while, the entropy in the universe continues to increase even as biological complexity on Earth, and perhaps at other sites in the universe, also increases. Evolution, like the sculpting of *David*, occurred without contradicting the 2nd law of thermodynamics.

Misunderstandings about Apes and Human Evolution

Some disbelievers in human evolution cite as evidence for their position the fact that no gorilla, gibbon, chimpanzee, or orangutan has ever birthed a human being. This is certainly true. But evolutionary theory does not claim that this ever happened or ever will happen. What evolutionary theory claims, and what the fossil record and DNA analyses show to be true, is that members of the great ape family (*Hominidae*) and other primates share common ancestors (see Figure 13.2). Hominids include *Homo sapiens,* the other four great ape groups named above, and numerous extinct species. So you and I and our parents *are* all apes. Our closest living, nonhuman relative is the chimpanzee. The common ancestor for humans and chimps lived about six million years ago and was neither a chimpanzee nor a human being. It was a species of the *Hominidae*; that is, it was an ape species which is now extinct.

Another misunderstanding about apes and human evolution is the notion of a "missing link." Persons doubting that Darwinian evolution applies to humans often ask, "Where is the missing link?" By *missing link*, evolution's doubters usually mean an undiscovered, fossilized being with physical characteristics midway between apes and modern humans. Evolutionary biology has three responses to the missing link question:

(1) Modern humans are apes.

(2) Evolutionary theory does not claim that there ever was an animal that was half modern gorilla or chimpanzee and half modern human, only that humans and nonhuman apes had a common ancestor different from any modern forms of apes.

(3) The human ancestral tree has a multitude of well-documented members, several of which Shawn Jacobsen describes in Chapter 6. Fossilized skeletons of hundreds of humankind's extinct, bipedal cousins, many from Africa, span a period from Ardi (*Ardipithecus ramidus*) 4.4 million years ago[9] up to the cave artists of Spain and France who began painting just 32,000 years ago.

Atheism or Rejection of Scripture Not Required

Many persons who accept biological evolution as an established fact also believe in the existence of a supernatural being and use sacred scripture as a moral guide for living. Both scientists and clergypersons have written about

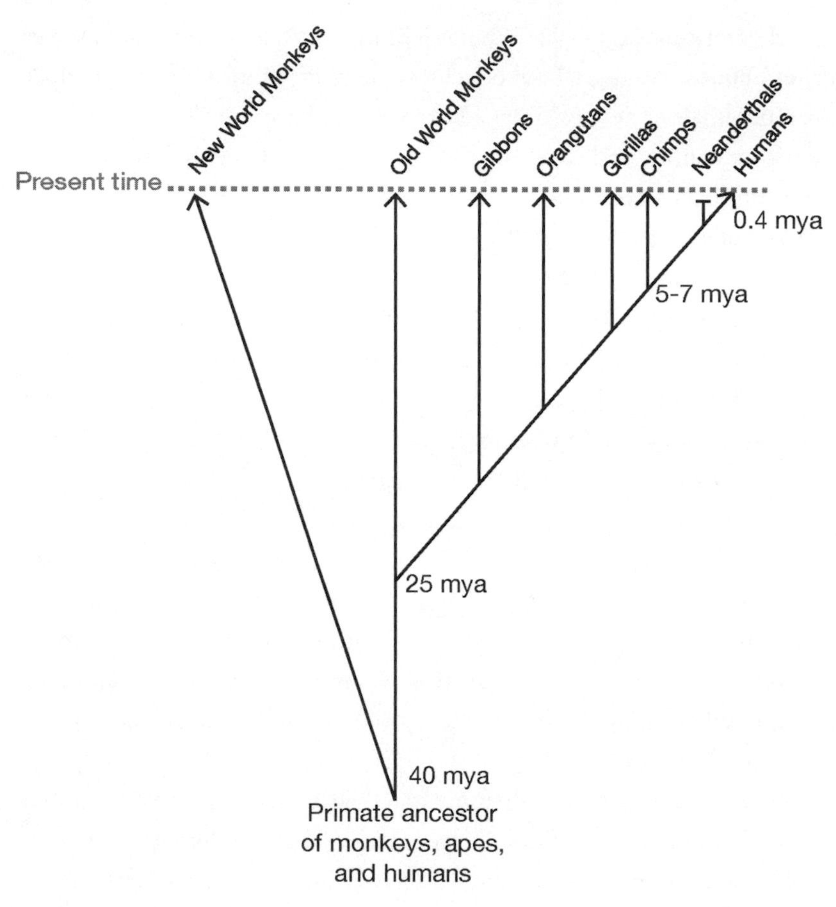

FIGURE 13.2

Higher primate phylogenetic tree showing that chimpanzees are more closely related to us than other living primates, and that the now extinct Neanderthals coexisted with us for a while and were closely related to us.

the compatibility of evolution and religious belief. Evolutionary biologist, biology textbook author, and professed Christian Kenneth Miller writes elegantly and lucidly about reason and faith in his 1999 book, *Finding Darwin's God: A Scientist's Search for Common Ground Between God and Evolution*. National Institutes of Health director and former leader of the publicly funded Human Genome Project, Francis Collins, does similarly in his 2006 book, *The Language of God: A Scientist Presents Evidence for Belief*.

And theoretical atomic physicist and Anglican priest John Polkinghorne writes about a compatible relationship between faith and science in his 1998 book, *Belief in God in an Age of Science*.

Most mainline denomonations have official statements about evolution. For example, a 1969 "Evolution Statement" by the Presbyterian Church (USA) reads in part:

> . . . Nowhere is the process by which God made, created or formed man set out in scientific terms. A description of this process in its physical aspects is a matter of natural science. The *Bible* is not a book of science . . ."[10]

The governing body of the United Methodist Church passed three evolution-friendly petitions in 2008, and one of these specifically opposes the introduction of faith-based views such as creationism and intelligent design into the science curriculum of public schools.[11]

The position of the Roman Catholic Church on biological evolution is more difficult to pin down. The Vatican seems comfortable leaving the evolutionary development of physical entities, including human beings, to science so long as the human soul remains under the purview of God.[12] Finally, the Clergy Letter Project, an open letter signed by more than 12,000 Christian clergy in the Unites States, states: "Religious truth is of a different order from scientific truth. Its purpose is not to convey scientific information but to transform hearts." On evolution, the Clergy Letter reads: "We believe that the theory of evolution is a foundational scientific truth, one that has stood up to rigorous scrutiny and upon which much of human knowledge and achievement rests. To reject this truth or to treat it as 'one theory among others' is to deliberately embrace scientific ignorance and transmit such ignorance to our children."[13]

Testimonies from the above theistic scientists and many others and statements from clergy from diverse faiths provide strong evidence that reasonable minds find compatibility between the findings of science and religious faith. Acceptance of the theory of evolution requires neither atheism nor rejection of the wisdom found in sacred scripture.

Why Knowledge about Evolution Matters

Denying Darwin creates evolution illiteracy. The results of Darwin denial are the same whether the denial results from actively avoiding learning, or allowing others to learn, about biological evolution; grows from the deliberate misrepresentation of evolution; or stems passively from lack of knowledge. Regardless of its source, Darwin denial in the United States has reached the point where it threatens the country's leadership in science, imperils the planet's ecosystems, handicaps future biomedical research, undermines the positive roles of religion in society, and fetters the human spirit.

"Nothing in biology makes sense except in the light of evolution." Biologist Theodosius Dobzhansky used this declaration as the title for his article written for biology teachers in 1973. The statement is as true today as it was then. Principles of biological evolution are foundational for every subdiscipline of biological science. Anatomy, biochemistry, plant and animal physiology, population biology, plant and animal ecology, cell biology, genomics, microbiology, neurobiology, nutrition, entomology, botany, ethnology, proteomics, genetics, molecular biology, parasitology, protozoology, marine biology, ornithology, mammalogy, and herpetology are all intimately integrated due to evolution. Discoveries in one subdiscipline guide research leading to new discoveries in other disciplines because plants, animals, and microbes have all coevolved. The DNA, biochemistry, physiology, behavior, and anatomy within individual organisms are also intimately integrated due to their coevolution.

An orchid and a moth provide a fantastic example of a prediction based on coevolution. Noting a pool of nectar deep inside the blossom of an orchid nearly a foot from the flower's opening, Darwin predicted the existence of a yet undiscovered moth with a proboscis long enough to reach the nectar and pollinate the plant. Sure enough, a half century later, a hawk moth with a proboscis long enough to do the job was discovered living in the same region of Madagascar as the orchid. The bottom line is that all of biology makes sense in the context of evolution, and anybody doing biological research without taking into account the evolutionary history of his or her objects of study is severely handicapped.

Modern Medicine Depends upon Evolution Theory

From the development of new drugs and surgical procedures to the annual development of flu vaccines, evolutionary principles guide biomedical research. Animals used in biomedical research have physiologies and biochemistries similar to that of humans. The reason mice and monkeys, rather than jellyfish and clams, are used by biomedical researchers is that they are closer to us on the evolutionary tree of life and therefore more likely to respond like humans to various drugs and procedures.

Evolution is why new flu vaccines must be grown each year. Flu-causing viruses do not stand still. As they reproduce inside host organisms, they rapidly evolve. After just one year our immune system cannot even recognize them as descendants of the bug that made us sick (or that we were immunized against) during the last flu season. The human immunodeficiency virus (HIV) that causes AIDS is another disease agent that evolves so rapidly that researchers have been stymied in their attempts to develop an effective vaccine against it. Bacteria that cause diseases like tuberculosis also evolve. They are especially good at evolving to become resistant to antibiotics we throw at them. The over-prescription and over-use of antibiotics is unwise because it gives bacteria the opportunity to evolve into resistant strains. The less prevalent antibiotics are in the environment and in our bodies, the more effective they will be in killing disease-causing bacteria when called upon to do so. Successful biomedical research for the cure and prevention of disease depends upon the understanding of the principles of biological evolution. The new field of evolutionary medicine is rapidly gaining momentum with journals, books, and medical school programs devoted to burgeoning medical applications of evolutionary biology.[14]

Environmental Protection Needs Evolution Theory

The lives of plants, animals, and microbes are interdependent within their communities and ecosystems. When human "progress" endangers or destroys just a single species of predator, prey, scavenger, decomposer (worms and microbes that recycle nutrients in dead plants and animals), pollinator, or photosynthetic producer, balance within the ecosystem is imperiled and the whole living pyramid may disintegrate. The interdependency living things have among each other and with particular physical environments, such as climate and nutrient availability, is a product of evolution. Understanding the principles of evolution helps us to determine when a population of plants or animals is so low that its continued existence and the health of the ecosystem are endangered. Principles of evolution also teach us about our own dependencies upon other living things and the physical environment. In fact, how well we understand and respect evolutionary principles may ultimately determine how long our species survives on the planet.

BIOTECHNOLOGY NEEDS EVOLUTIONARY BIOLOGY

Ten thousand years ago humans began changing the genetic make-up of plants and animals. Early agriculturalists unknowingly used principles of evolution by natural selection to domesticate wheat. Year after year they replanted the grain that they harvested the season before. In so doing, they selected seed stock for plants with grain heads able to hold on to their seeds rather than letting them fly away into the wind. Before long, humans were selectively breeding plant and animal variants to obtain offspring better able to meet human needs. We still use these slow and laborious methods to obtain new and better agricultural plants and animals. But since the 1980s humans also have been able to leapfrog over the slow steps of evolution by human selection and directly manipulate the DNA of virtually any organism on Earth.

Genetic engineering has produced corn plants with bacterial genes that make them insect-resistant, bean plants with a daffodil gene that makes them herbicide-resistant, and bacteria that produce human insulin and growth hormone. In 2009 researchers created designer bacteria that produce lycopene, an antioxidant in tomatoes that promotes heart health. Employing principles of biological evolution, the researchers selected and grew variants of the lycopene-producing bacteria that were most efficient at producing lycopene.[15] Genetic engineers, in collaboration with electrical and computer science engineers, have spawned a new discipline called synbio (synthetic biology). Synbio aims to create brand new forms of life, perhaps even multicelled organisms, to perform specific tasks. One can imagine many good results that could come from synbio technology, including cleaning toxic waste dumps, removing greenhouse gasses from the atmosphere, producing new food and energy sources, and fighting diseases in new and creative ways. But genetic engineering and synbio, done in the absence of knowledge about evolution and ecology, could also cause great harm to the planet.

The wisdom of natural selection acting over four billion years has produced our present biosphere with its countless and exquisitely balanced interdependent components. What wisdom will guide the introduction of new life forms, completely foreign to anything the Earth has ever seen, into the biosphere? If it is the "wisdom" of the marketplace and the "wisdom" of human desire for comfort and entertainment, rather than a deep understanding of principles of ecology and biological evolution, the outcome for evolution's splendid array of living things may be disastrous. Foreign, human-designed organisms could out-compete some of nature's life forms and send entire ecosystems out of balance. Ethicists Joachim Boldt and Oliver Müller write that "seen from the perspective of synthetic biology, nature is a blank space to be filled with whatever we wish."[16] Nature could easily become a blank space to be filled with our own clever constructions rather than a space already occupied with organisms to be appreciated, understood, and preserved. Effective environmental preservation depends upon knowledge about biological evolution.

"Equal Time" Undermines Both Science and Religion

Attempts to mandate Bible-based creationist views, or the more subtly veiled religion-based intelligent design views, of the origin of species to be taught alongside biological evolution in U.S. public school science classrooms have gone to court many times.[17] At issue is whether teaching creationism or intelligent design as credible alternatives to evolutionary theory violates the First Amendment of the U.S. Constitution, which prohibits laws respecting an establishment of religion. Two well-known federal court cases are *McLean v. Arkansas Board of Education* (1982) and *Kitzmiller et al. v. Dover* (2005). In the former, the court ruled that an Arkansas statute requiring public schools to give balanced treatment to "creation-science" and evolution-science violated the Establishment Clause of the First Amendment. In the latter case, U.S. District Court Judge John E. Jones III ruled similarly for the teaching of intelligent design in the Dover (Pennsylvania) Area School District.

When parents opt for private schooling or home schooling to shield children from the theory of evolution, or to teach religion-based accounts of creation as though they were science-based alternatives to evolutionary theory, they undermine children's comprehension of what science is and fetter later careers in science. Confusion and conflation of two very different ways of knowing—knowing through revelation and symbolic language versus knowing through the self-correcting, hypothetico-deductive method of science—hampers students' development in both realms of knowledge.

Denial Undermines Positive Influences of Religion

When a child's religious training teaches that evolutionary theory lacks good scientific evidence and rigor, that evolution is an atheistic scheme to destroy belief in God, that believing in evolution precludes religious faith, that the Earth is only a few thousand years old, or that a literal reading of sacred scripture is a reliable account of how the physical universe actually came into being, the stage is set for a great disillusionment with religion down the road. For when that child begins reading and thinking for himself, gets out into the world and converses with honest and intelligent persons of different educational backgrounds, or takes a good biology course in high school or at the university, he will discover that what he was taught earlier about evolution and scripture is untrue. Any one of a multitude of discoveries about the world around him could jeopardize all of the religious teachings he received as a child. For example, learning the true meaning of the term *scientific theory*, how geologists date the Earth at 4.6 billion years, about *Tiktaalik*, feathered dinosaurs, or extinct species of the genus *Homo*, meeting a religious evolutionary biologist, or realizing that the two accounts of creation in the Bible's Book of Genesis are mutually exclusive may create a crisis of faith. The result may be to discard religion as deceptive and unworthy of further attention. But religion can nurture courage and hope, kindness and compassion, cooperation and fellowship, and the ability to envision peace during times of war. To jeopardize these and other virtues of religion by railing dogmatically against evolution does violence against the future personal lives of children and society as a whole.

Galileo cited Cardinal Baronius (1598) when in 1615 he wrote to his friend, the Grand Duchess Christina of Tuscany, that scripture teaches "how to go to heaven, not how the heavens go."[18] Galileo was writing about the conflict between his new telescopic discoveries and the Roman Catholic Church's insistence that Aristotle's Earth-centered universe was correct because it agreed with literal readings of scripture. In 1633 Galileo was arrested by the Holy Office of the Inquisition and convicted of teaching that the Earth orbits the sun. Nearly 350 years later, the Church officially forgave Galileo. Perhaps those who teach that evolution has not and does not occur can learn from this episode in the history of Christianity and consider the possibility that scripture teaches us how to live life, not how life came to be.

Pitfalls of Shielding Children from Knowledge

From anatomy and physiology to behavior, humans share many characteristics with other animals. But humans are unique in their abilities to devise language, use speech to communicate abstract concepts, and engage in the reasoning that produces nuclear power plants, modern biotechnologies, and books about moral philosophy. Normal brain development produces children who are curious about themselves and the rest of the universe. Human curiosity, in partnership with reason, education, and communication, fuels humankind's cultural advancements in art, science, technology, and social living in the global community. Purposefully denying students the opportunity to learn what science has to say about the interrelatedness of all components of the biological world, confusing them about the nature of science, and discouraging their questions or skepticism about dogmatic teachings are hostile to the human spirit. Each of these hindrances to normal human development is especially egregious when carried out by persons who themselves have not made the effort to learn what evolutionary theory actually is, and what it is not.

Plato's Cave and Learning about Biological Evolution

Consider now a simple exercise to aid students and teachers in discussing evolution in the classroom and to help anybody learn or teach about evolution. Plato (428/427–348/347 BCE) can help us, not because he knew about biological evolution, but because he knew about human nature. In Book VII of Plato's *Republic*, Socrates famously describes what can happen to any of us when challenged by new ideas or when introducing others to new ideas.[19] Allow me to paraphrase and take some liberty in interpreting Plato's Allegory of the Cave for the advantage of students and teachers of evolution.

One sunny day, a person enters a cave to find inside a group of "prisoners" tied in chairs facing one wall of the cave. They cannot even turn their heads from side to side and have spent their entire lives like this. Moving before them on the wall of the cave are dark, indistinct images of animals, people, and implements of all kinds. The intruder examines the situation and discovers a line of puppets and other objects hanging from a stone ledge behind the prisoners; behind the puppets burns a fire whose light is interrupted by the puppets before it illuminates the wall facing the prisoners. So, for the prisoners' entire lives, reality has been the shadows cast by puppets. Recognizing a teachable moment, the intruder unties one of the prisoners, shows her the puppets and the fire, and then leads her up out of the cave into the sunlight. Entrance into the above-ground world of sunlight is painful to her eyes. But after some time, the former prisoner adapts to the light and soaks up the wonders and beauty of the sunlit world. Anxious to tell her cave-bound family members and friends what she has discovered, she returns to the cave. But to her surprise and dismay, her words are met with hostility and ridicule. Her family and friends believe she has lost her mind.

The pain we experience when moving from a dark place into the sunlight, and the hostility we receive from those living in darkness are risks of education. Pain and hostility are normal responses to new ideas that upset our previous world views. Caves are comfortable. They are neither too hot nor too cold; they remain predictably about the same all of the time. Each of us has some personal caves in which we live or to which we retreat. When new information makes us feel uncomfortable or angry, it may be revealing a personal cave to us. Our job then is two-fold: to decide whether our cave is based on something negative such as fear, prejudice, or ignorance; and to decide whether we will risk leaving the comfort of the cave to move toward the light of reason and education.

Since discovering Plato's Allegory of the Cave 25 years ago, I have used it regularly and successfully in university biology courses to aid in teaching and learning about controversial topics, including human cloning, human embryonic stem cell research, scientific origin of life studies, and biological evolution. Students who have read or heard Plato's cave allegory report feeling more open-minded and willing to challenge themselves with new ideas. Finally, it is as important for teachers to acknowledge that they have their personal caves as it is for them to encourage students to identify their own caves and begin stepping out of them.

Acknowledgment

Designer and photographer Janna Claire Sidwell created the illustrations for this chapter. ❧

Notes

1 Gallup. On Darwin's Birthday, Only 4 in 10 believe in Evolution. http://www.gallup.com/poll/114544/darwin-birthday-believe-evolution.aspx.

2 Time. Can You Believe in God and Evolution. http://www.time.com/time/magazine/article/0,9171,1090921-1,00.html.

3 Brown University. Case Histories of Speciation I&II. http://biomed.brown.edu/Courses/BIO48/23.Cases.html.

4 Barluenga, M., et al. 2006. "Sympatric Speciation in Nicaraguan Crater Lake Cichlid Fish." *Nature* 439: 719–23; Carson, H.L., 1982. "Evolution of Drosophila on the Newer Hawaiian Volcanoes." *Heredity* 48: 1–25; Witte, F., et al., 2008. "Major Morphological Changes in a Lake Victoria Cichlid Fish within Two Decades." *Biological Journal of the Linnean Society* 94: 41–52. An excellent, short description of examples of speciation written by a biology professor for his evolutionary biology course at Brown University uses these sources.

5 Ridley, M. 1996. Macroevolutionary Change. In *Evolution,* 2nd ed. Cambridge: Blackwell Science, Inc., 582–609.

6 Dawkins, R., 1996. The "Alabama Insert:" A Study in Ignorance and Dishonesty. In *Charles Darwin: A Celebration of his Life and Legacy*. Montgomery: NewSouth Books.

7 Gee, H., et al. 2009. Land-Living Ancestors of Whales. 15 Evolutionary Gems.

http://www.sesbe.org/sites/sesbe.org/files/file/EVOLUTIONARY_GEMS.pdf.

8 Reuters. 2009. China Finds Bird-Like Dinosaur with Four Wings. http://www.reuters.com/article/idUSTRE58R1ST20090928.

9 University of Chicago. Meet *Tiktaalik roseae*: An Extraordinary Fossil Fish. http://tiktaalik.uchicago.edu/meetTik.html; Gee, H., et al. 2009. From Water to Land. 15 Evolutionary Gems. http://www.sesbe.org/sites/sesbe.org/files/file/EVOLUTIONARY_GEMS.pdf; Gibbons, A. 2009. "Breakthrough of the Year: *Ardipithecus ramidus.*" *Science* 326: 1598–9.

10 Presbyterian Church USA. Theology and Worship; Evolution Statement. http://www.pcusa.org/theologyandworship/science/evolution.htm.

11 United Methodist Portal. Methodism Supports Teaching of Evolution. http://www.umportal.org/article.asp?id=3869.

12 Catholic Answers. Adam, Eve, and Evolution. http://www.catholic.com/library/Adam_Eve_and_Evolution.asp.

13 Clergy Letter Project. Statement on Evolution. http://www.clergyletterproject.net/.

14 Pennisi, E. 2009. "Darwin Applies to Medical School." *Science* 324: 162–3.

15 Wang, H., et al. 2009. "Programming Cells by Multiplex Genome Engineering and Accelerated Evolution." *Nature* 460: 894–8.

16 Boldt, J. and O. Müller. 2008. "Newtons of the Leaves of Grass." *Nature Biotechnology* 26: 387–8. The authors argue for an extended dialogue among synthetic biologists to develop a code of ethics for their discipline that reflects an understanding of how synbio activities impact society and nature as a whole.

17 Matsumura, M. and L. Mead. 2007. Ten Major Court Cases about Evolution and Creationism. National Center for Science Education. http://ncse.com/taking-action/ten-major-court-cases-evolution-creationism.

18 Galilei, G., 1615. Letter to the Grand Duchess Christina of Tuscany. In *Discoveries and Opinions of Galileo.* New York: Random House, Inc, 1957.

19 Plato. ca. 380BC. *The Collected Dialogues, Republic Book VII.* eds. E. Hamilton and H. Cairns. Princeton: Princeton University Press, 1989, 514–18.

Contributors

JONATHAN ARMBRUSTER is an Alumni Professor in the Department of Biological Sciences at Auburn University. He received his BS and PhD from the University of Illinois in the Department of Ecology, Ethology, and Evolution. He teaches comparative vertebrate anatomy and evolution and systematics to undergraduate and graduate students. His research is primarily on the evolution and taxonomy of the suckermouth armored catfishes of South America. He is also curator of Auburn University's Biodiversity Learning Center's fish collection. Dr. Armbruster also played the role of Darwin throughout Auburn's celebration of Darwin in 2009.

LEWIS BARKER is a Professor of Psychology and coordinator of an interdisciplinary program, the Human Odyssey, at Auburn University. He earned an AB in Psychology from Occidental College (Los Angeles) and an MS and PhD in Psychobiology from Florida State University. Before coming to Auburn, he taught at from Baylor University where he was a Professor of Psychology and Neuroscience. His teaching and research interests have included human and animal learning and memory, and evolutionary psychology.

GUY V. BECKWITH is Associate Professor of History at Auburn University. He earned a BA in English from the University of California at Santa Cruz, an MA in English from the University of California, Santa Barbara, and an interdisciplinary PhD focused on the history and philosophy of technology, also from Santa Barbara. He teaches courses in Technology and Civilization for undergraduates, and a seminar in the Theory of History for PhD candidates. His research interests include the uses of technological imagery in the western literary tradition, and the cultural role of technology in the ancient world.

JAMES T. BRADLEY is the W. Kelly Mosley Professor Emeritus of Science and Humanities and Professor Emeritus of Biology in the Department of Biological Sciences at Auburn University. He earned a BS in Biochemistry from the University of Wisconsin (Madison) and a PhD in Developmental Biology from the University of Washington. He teaches cell biology and bioethics for pre-professional undergraduate students and graduate students in the life sciences. His research interests have included the biochemistry of yolk formation in fish and insect eggs and the endocrine control of insect reproduction.

RICHARD DAWKINS is a British ethologist, evolutionary biologist and author. He is an emeritus fellow of New College, Oxford, and was the University of Oxford's Professor for Public Understanding of Science from 1995 until 2008. Dawkins came to prominence with his 1976 book *The Selfish Gene*, followed by *The Extended Phenotype* (1982), *Climbing Mount Improbable* (1996), and *The Greatest Show on Earth: The Evidence for Evolution* (2009). His most recent work is *The Magic of Reality: How We Know What's Really True* (2011).

GERARD ELFSTROM is Professor of Philosophy at Auburn University. He earned his BA from Cornell College and an MA and PhD from Emory University, all in philosophy. Though he has taught a variety of philosophy courses, for the past several years, he has primarily offered an introduction to logic. He has done research in applied ethics, the ethics of international relations, and the philosophy of science.

DEBBIE FOLKERTS is Assistant Professor of Biology in the Department of Biological Sciences at Auburn University. She earned a BS in Biology and a MS in Entomology from Auburn University and a PhD in Entomology from the University of Georgia. She teaches a variety of classes on organismal biology at freshman, senior, and graduate levels. Her research interests include carnivorous plant biology, plant-animal interactions, and spider ecology. She dedicates her chapter to the memory of her late husband, George W. Folkerts, who opened her eyes and the eyes of countless students to the wonders of nature, the joys of studying biology, and an understanding of life in the context of evolution through natural selection.

KENNETH M. HALANYCH is an Alumni Professor in the Department of Biological Sciences at Auburn University. He also serves as the Marine Biology Liaison for the University. He earned a BS in Biology from Wake Forest University and a PhD in Zoology from the University of Texas. He teaches invertebrate biology and marine biology for students in the life sciences. His research interests included the evolution of major animal lineages, evolution of Antarctic marine invertebrates, and deep-sea biology.

SHAWN JACOBSEN IS the field laboratory coordinator for the Department of Biological Sciences at Auburn University. He received a BS in Wildlife and Fisheries Sciences from Texas A&M University and an MS in Zoology and Wildlife Sciences from Auburn University. He coordinates and teaches ecology and vertebrate biodiversity laboratories. He is interested in all aspects of field biology, especially herpetology, and enjoys writing science fiction.

JEFFREY S. KATZ is an Alumni Professor in the Department of Psychology at Auburn University. He earned a BA in Psychology from Ithaca College and an MS and PhD in Experimental Psychology from Tufts University. He teaches undergraduate and graduate classes in Cognitive Neursoscience, Cognitive Psychology, Comparative Cognition, and Sensation and Perception. His research focus is in the area of comparative cognition and has been funded by National Institutes of Health and National Science Foundation. Current projects involve abstract-concept learning, change detection, memory processes, and neuroimaging.

DAVID T. KING JR. is Professor of Geology in the Department of Geology and Geography at Auburn University. He is also director of the Concepts of Science program at Auburn. He earned a BS in geology from the University of Louisiana-Monroe, an MS in geology from the University of Houston, and a PhD in geology from the University of Missouri-Columbia. He teaches integrated science, introductory geology, and sedimentary and impact geology courses. His research interests include the effects of extraterrestrial impact upon Earth history and the rock record.

JAY LAMAR IS the Director of Special Programs for the Office of the Provost and University Libraries at Auburn University. She serves directly with the Office of Undergraduate Studies on special lectures and the common reading program and with the Department of Archives and Special Collections in the Ralph B. Draughon Library. For more than two decades she was director of the Caroline Marshall Draughon Center for the Arts & Humanities, where she helped forge partnerships between the university and the community. She is co-editor of *The Remembered Gate: Memoirs by Alabama Writers*.

JOHN F. MAGNOTTI is a doctoral student at Auburn University in Experimental Psychology. He earned a BS in Psychology and Computer Science from James Madison University, and an MS in Experimental Psychology from Auburn University. He has assisted in teaching graduate and undergraduate applied statistics, as well as general psychology. His research interests include basic issues in the comparative study of visual perception and the use of neuroimaging techniques to study fundamental processes of human memory.

ANTHONY MOSS is Associate Professor of Biology in the Department of Biological Sciences at Auburn University. He earned a BA in Cell Biology from Johns Hopkins University and a PhD in Marine Cell Biology and Physiology from Boston University Marine Program. He teaches undergraduate Honors Biology, undergraduate and graduate Cell Biology, Wound Repair, and Light Microscopy. His training was originally in cellular physiology and biochemistry, and he currently studies ctenophore and jellyfish structure and function, as well as the marine microbes associated with jellyfish.

Kelly A. Schmidtke has worked as an instructor and/or researcher in at various institutions in the USA and UK, with economic, philosophy, psychology, and veterinary departments. She earned a BA from the University of Minnesota-Twin Cities and an MS and PhD from Auburn University. Her teaching and research interests include animal models of human behavior and evolutionary psychology.

Giovanna Summerfield is Associate Dean for Educational Affairs for the Auburn University College of Liberal Arts. She is also an associate professor in the Department of Foreign Languages and Literatures at Auburn University. She received a BA in Government and Politics from the University of Maryland, College Park, and a PhD in Romance Languages and Literatures from the University of Florida. She teaches Italian and French at the undergraduate and graduate level. Her research and teaching interests focus on the long 18th-century (1660–1830) French and Italian literature (emphasis on Sicilian writers), religious and philosophical movements, women's studies, and material culture. She is also a published poet and short-story writer.

Anthony A. Wright has taught since 1982 at the University of Texas Medical School as a Professor in the Neurobiology and Anatomy department. He earned a BA in Psychology from Stanford University and an MA and PhD in Psychology from Columbia University. He and collaborators have compared list memory processing, visual working memory, and relational versus item-specific learning in pigeons, capuchin monkeys, rhesus monkeys, and humans including patient groups. His work has been funded by the National Institutes of Health and the National Science Foundation.

Index